From Hoofbeats
to Dogsteps

A Life of Listening to and
Learning from Animals

Rachel Page Elliott

Publishing

Wenatchee, Washington U.S.A.

From Hoofbeats to Dogsteps
A Life of Listening to and Learning from Animals
Rachel Page Elliott

Dogwise Publishing
A Division of Direct Book Service, Inc.
403 South Mission Street, Wenatchee, Washington 98801
1-509-663-9115, 1-800-776-2665
www.dogwisepublishing.com / info@dogwisepublishing.com

Photos: Elliott family albums, Jane Donahue, Chris O'Conner, and Hal Ungerleider

Limits of Liability and Disclaimer of Warranty:
The author and publisher shall not be liable in the event of incidental or consequential damages in connection with, or arising out of, the furnishing, performance, or use of the instructions and suggestions contained in this book.

Library of Congress Cataloging-in-Publication data
    Elliott, Rachel Page.
    From hoofbeats to dogsteps : a life of listening to and learning from animals / Rachel Page Elliott.
    p. cm.
    ISBN 978-1-929242-60-3
    1. Elliott, Rachel Page. 2. Pet owners--Masschusetts--Biography. 3. Pets--Masschusetts--Anecdotes. 4. Domestic animals--Masschusetts--Anecdotes. 5. Human-animal relationships--Masschusetts--Anecdotes. 6. Farm life--Masschusetts--Anecdotes. I. Title.
    SF411.45.E45A3 2009
    636.088'7--dc22
    [B]
    2008031418

ISBN: 978-1-929242-60-3

Printed in the U.S.A.

More praise for *From Hoofbeats to Dogsteps*

Rachel Page Elliott has been the most notable spokesperson for structure and gait for the past five decades. We met at one of her lectures and decided that we could join efforts and offer a two day seminar that featured several topics. I would speak on breeding to produce a better dog and Pagey, would through her film presentation and lecture, explain the principles of structure, gait and its importance to breeders. She was gifted at taking what can be a complicated topic and making it understandable. These seminars were especially popular because breeders were hungry for information. Pagey was prepared for the challenge and would explain why shoulders (lay-in, lay-back, length of upper arm etc) were so important to good movement. She clarified the expression "a good front" by explaining what it meant and then visually demonstrated the important principles through lecture and film…

Pagey has, over the past 50 years, served as the beacon of light for those interested in understanding the principles of structure and motion…Her book and DVD on these subjects are a must read for every dog breeder.

Carmen Battaglia, author of *Breeding Better Dogs* book and DVD, AKC Board Member and Judge

This book gives us a glimpse into the life of a woman whose life-long love of animals led her to become one of the world's most widely respected authorities on canine structure and movement.

Anne Shannon, Past President, Golden Retriever Club of America

Pagey has been a cherished friend of jigsaw puzzlers everywhere for decades, as well as a stunningly talented puzzle cutter in her own right.

Anne D. Williams, Professor at Bates College and author of *The Jigsaw Puzzle: Piecing Together A History*

Family Retrospective

As my brother, David, sister, Betty and I have sat with our mother over the last ten months working on this memoir we have had the joyful opportunity of reliving our own lives. Poring over scrapbooks, reading old letters, looking at news clippings and sharing memories has been a treasured experience. We all knew "Pagey" was a golden lady but until now we did not know how important she had become to dog and animal lovers around the world. She was just our mother who traveled in and out of River Road Farm on canine and equine missions as easily as most mothers would go to the grocery store. All along she was a writer, researcher, speaker and leader as this wonderful memoir describes. She was also our role model since we never knew she was doing anything out of the ordinary.

Our living room was always full of visitors with dogs and animals of every size and shape, now with familiar names in this book. Most of the time, Mother was on the phone, at her typewriter, holding a meeting or somewhere with her four-footed friends. Until now, we never knew she was in the vanguard of so many important projects. She was just "Mother" who loved Dad and gave us all a very remarkable childhood. We are glad to have shared her with so many people from around the globe and we extend our gratitude to everyone who has touched her life in so many ways all of these years.

Ruth Elliott Holmes
July 2008

CONTENTS

Dedication

To my late husband, Dr. Mark Elliott, who joined me on this memorable journey.

ACKNOWLEDGMENTS

I have come to realize that one of the greatest challenges of writing my memoir is trying to adequately thank the countless people and organizations who have contributed so greatly to my lifelong quest. A memoir is not just about the author. It is about the people who made a life worth writing about.

The support of my family has been crucial during the last year of trial and error, trying to keep the script interesting and meaningful. My son, David, and daughter, Ruth have worked tirelessly on the final manuscript, searching out scrapbook photographs and polishing the project to make it ready for the publisher. My granddaughters, Cindy and Sarah, helped type, format and layout the original manuscript. My lawyer grandson, Nicholas, has helped with the legalities of the endeavor.

Through it all, my oldest daughter, Betty, her husband, Maris, and daughter, Rachel, continue to keep River Road Farm running while I enjoy the best of it. For fifty-five years, always nearby with an encouraging word and a helping hand, is my special friend, Alberta White. To my entire extended family, I express my loving thanks.

My special thanks to Anne Shannon for her Foreword. My appreciation goes to Chris O'Connor for taking one of the cover photographs and those who kindly wrote testimonials, Nini Bloch and Dr. Chris Zink. My dear friends Dottie Norwood, Jan Kennedy, Mark and Jan Bramhall listened endlessly to my tales while Donna McKibben Cutler skillfully scanned my old photographs to accompany the stories. I also wish to thank development editor, Richard Higgins, without whose guidance I might never have started this task. He offered structure and form to the text, and kept my side-trips down memory lane from straying too far afield.

I am tempted to write several pages listing the names of those who have contributed so greatly to my studies, my writing projects, my travels, and my life, a life filled with memorable people and animals. Instead, lest I offend someone by inadvertent omission, let me say that if you know me you have contributed to this wonderful journey and I am grateful. Even those with whom I have had differing opinions have contributed since they have spurred me ahead in my research.

So many friends in so many circles, not just limited to dogs, horses and puzzles, have touched my life. I express my heartfelt gratefulness to each and every one of you.

To all of my friends, near and far, known and unknown, my gratitude runs deep for your interest, encouragement and support for an understanding and commitment to the betterment of the animals that so enrich our lives.

River Road Farm
291 River Road
Carlisle, MA 01741

FOREWORD

In the early 1970s, I was one of a group of dog lovers who poured into a lecture hall in Greeley, Colorado—eager to hear the author of a recently published book entitled *Dogsteps* present her film by the same name. A tall, distinguished woman with a firm, commanding voice took her position next to her projector and led us on a journey that, for many of us, has never ended. Over the years, Rachel Page Elliott, or Pagey as she is known to all, went on to lecture on several continents, patiently educating and enlightening those who desired to learn.

This book gives us a glimpse into the life of a woman whose life-long love of animals led her to become one of the world's most widely respected authorities on canine structure and movement.

Starting with her observations of family pets, horses and working dogs, Pagey soon began filming hundreds of dogs. Her inquisitive mind and critical study of these films soon led her to question many of the widely held beliefs and theories of the time. She eventually employed a technique known as "cineradiography"—a methodology for x-raying moving animals—that conclusively proved that the long held notion of the "45-degree layback of the shoulder" was mechanically impossible.

Although Pagey's work has aided the cause of all breeds, we who have Golden Retrievers have been incredibly blessed that Pagey chose our breed and our parent club as her very own. In 1945, Pagey registered her kennel name "Featherquest" with the American Kennel Club, establishing it as one of the oldest Golden Retriever kennels in the country.

Pagey's name has been emblazoned on the history of our breed for over 60 years. Indeed, it was Pagey, who in 1982, first identified and located the ruins of Guisachan, Lord Tweedmouth's family hunting estate in the Scottish Highlands, and recognized it as the birthplace of the Golden Retriever breed.

As determined as we may be to put Pagey on a pedestal for her graciousness, intelligence and leadership, Pagey prefers to stand beside us, helping us to learn and teaching us to question. In 1987, Dr. Quentin LaHam, a noted lecturer on canine structure, introduced Pagey at a lecture for Ladies Dog Club by saying, "When the history of dogs in the 20th century is written, the name at the top of the first page should be that of Rachel Page Elliott." Everyone who knows her would concur!

Yet, as readers of this book of memories will soon learn, Pagey's personal charm transcends the scholarship for which she is best known. Though her life achievements are impressive and her expertise undisputed, it is her unparalleled character and grace that cause her to stand out as such an exceptional personality.

A trip to Pagey's beloved River Road Farm is as magical an experience as can be imagined! The phrase "never a dull moment" takes on new meaning when confronted by the myriad of activities that make up a day. Pagey quickly draws you into her world—whether making marmalade, hand cutting intricate jigsaw puzzle pieces, discussing dogs of every variety, walking the pastures alongside the Concord River, or harnessing up a Connemara pony and taking a buggy ride. And the visitors! Friends, neighbors and overseas travelers stop by many times daily. All are met with a cup of tea and some variety of sweets. If the timing is right, a touch of sherry may be brought out to celebrate. And then there are the animals—always the animals! A variety of cats grudgingly share their chairs, Goldens and Corgis have their places, and the beautiful Connemara ponies gather at the fence to give a nudge in greeting.

It is fitting to end this foreword with the verse Pagey never fails to recite when surrounded by her many friends and family. I'm sure those who know her will hear her distinctive voice in these words which so completely reflect her humility and graciousness.

It is my joy in life to find
At every turning of the road,
The strong arm of a comrade kind
To help me onward with my load.

But since I have no gold to give
And love alone must make amends
My only prayer is while I live
God make me worthy of my friends.
                                        —Anonymous

Anne Shannon,
Past President, Golden Retriever Club of America

Chapter 1

THE MEMORABLE MAJOR

Leaves from the great oak towering over the woodshed swirled around my feet as I struggled to secure the large green door lifted from its hinges by the fast-moving winter storm. To escape the wind and catch my breath I stepped inside the shed. Leaning against a well-traveled pony cart draped with harnesses, below a wall of saddles and bridles, I gazed around this dusty warm end of the 1701 farm house that I had grown to love over the last sixty five years. There I found myself in a treasure chest of memories.

As I peered around this old storage shed, my eyes fell upon the once sharp and shiny double-handled bucksaws that hung beside the duck decoys and wooden sleds. Plain and unadorned, these were my siblings' sleds. Unexpectedly, a flashback took me to a snowy day 90 years ago when our family's Shetland pony, Major, got stuck in a snow bank. I was four years old.

I remember winter mornings waking up to the soft plodding of hooves as horse-drawn plows cleared the fresh fallen snow from the sidewalks and roads. In those days, the early 1900s, the streets stayed packed with snow for weeks at a time. That's when we had fun hooking our sleds to the wide runners of the four-horse pungs, the large box-like sleighs used for delivering heavy loads of coal or huge cakes of ice, until the drivers chased us off.

Lexington Town Team plows snow, 1920s.

Ponies, Major and Jimmy, take the Webster children for a winter ride, 1925.

Even more fun was attaching five or six sleds, one behind the other, to long harness tracers behind our pony Major. He'd pull us over the snowy streets as we tossed snowballs to make him trot. Occasionally we would hitch him to his own small sleigh and sit on its comfortable cushioned seat.

One of my earliest childhood recollections of Major is his getting stuck when he rolled in a snow bank. In winter, he loved to paw and prance in the fresh snow. When I heard my sister's call for help one morning, I knew that our freedom-loving pony had escaped from his stall again. This time, however, his frolic did not last long, for as he frolicked in a huge snow drift left by a plow, he got caught upside down and could not move.

Fortunately the ice man, who happened to be delivering a large block of ice for the kitchen ice box, came to our rescue. He helped my father pull the pony out and onto his feet. Major shook the snow from his thick black coat and, without a glance of appreciation, galloped back to the barn.

I remember too, how even in winter, Major liked to stop at Lexington Green and reach up to dip his nose in the water trough at the base of the Minuteman statue, one of the most famous monuments to Revolutionary War soldiers. Drivers of the town's horse teams that often stopped there used to laugh at the pony's icy whiskers.

I grew up in Lexington, Massachusetts, the last of the six children of Hollis and Helen Webster, reared in the discipline of old New England values. My father loved stamp collecting and was a professor with a doctorate in romance languages from Harvard. My mother, born Helen Maria Noyes, went to Radcliffe—as would all four of her daughters. We learned from our family tree that some of our Noyes forebears were abolitionists who lived in Sag Harbor on the eastern tip of Long Island, close enough by sea to aid the 53 Africans who, in 1839, seized control of the slave ship Amistad in a bloody mutiny. The African men were captured in America and eventually acquitted by the United States Supreme Court. Our Noyes branch also started an academy in New Hampshire that accepted black students in 1835. The academy was later destroyed by arsonists. Another interesting ancestor of ours was John Rowe, the Boston maritime magnate and state representative who in 1784 donated the Sacred Cod, the wooden carving of the fabled fish that hangs above Rowe's former chamber in the Massachusetts State House. Rowe's Wharf still carries his name. It's always fun to dig into the family heritage. It was only as an adult that I learned we go directly back to the Rev. Dr. John Rogers, the fifth president of Harvard University.

Major and friends at Lexington Minuteman Monument.

We were fortunate to live in a large, white-clapboard house with a pillared front porch that looked out on Lexington Green, the site of the British massacre of our ready-in-a-minute militia of farmers and merchants who began the fighting in the American Revolution. Our house was not fancy, but it was roomy and comfortable. Maintaining it and feeding a family of eight, as well as sending five children to Radcliffe and Harvard, was sometimes a challenge on

Father's salary, but we managed. We were expected to do our home duties faithfully and take pride in our community, to appreciate school and respect teachers. Our home was a happy one because misbehavior was not lightly tolerated. This point was driven home the day Mother spanked me for swearing. I still recall the outburst. My older brother Albert helped water the horses but he didn't know how to ride. I didn't encourage him as I feared he would kick Major or jerk his mouth. When he'd been given a smart new cowboy suit, Mother offered him 10 cents to ride the pony down the street just to see him on horseback. I was mad!

It may surprise children today to know that we did without electric refrigerators, frozen meals, dishwashers, electric blankets, nylons and pantyhose. When I was quite young, the ball point pen had not been invented yet, somehow, we got by. I didn't know why I had to do a certain thing, but a blessing in those days was that I didn't have to know why, I simply did it. It was only years later, for example, that I appreciated the piano teacher who required me to practice my lessons one hour per day. My allotted time to practice on our family piano was each day from 6:15 to 7:15 am. Thus, even in winter, I began each day seated before the piano, often with a hot water bottle in my lap. This kept my fingers, if not nimble, at least moving up and down the keys. Now and then my brother Albert accompanied me on his flute.

We had several dogs including Saint Bernards, Cocker Spaniels, German Shepherds, Chows, collies and terriers and many other animals, all of whom added to our duties and contributed to our fun. Horses were my special love. I knew every horse in town by name and took it upon myself to upbraid the grocery delivery man for abusing his thin, overworked animal. The favorite of these horses was our family Shetland pony, Major. After surviving my three older sisters, Major fell to my care and was my closest companion during those early years of my life. He was my pony, and there were days when I cried into his mane as if he were my most understanding friend.

Major on the loose in front of our house, Lexington, early 1920s.

Our neighbors knew our pony well, patiently tolerating the nuisance of his getting loose and leaving hoof prints on their lawns or gardens. Once he even rolled on the surface of a freshly laid tennis court. My father quietly compensated the owner for the repair.

Major was always ready for tidbits to eat and never fussy about what was offered. I can see him now, his ears pricked toward our kitchen door as someone—is it me?—holds out a lump of freshly cooked horehound candy that had already hardened on a large silver spoon. Many years later I thought of Major fondly when I found that cherished spoon. How did I know? Because his bite marks on the spoon were still clearly visible.

I rode this sturdy pony everywhere, never with a saddle. He always seemed to know where I was going. A special treat for both of us was a visit to the local racetrack where we watched the trotters and pacers in training. Often we joined friends who had ponies and played "Cowboys and Indians" on the soft paths around the town playground. Riding was not allowed on Lexington Green but it was a comfortable place for pony feet. At quiet moments I would canter him alone along its far side, hoping not to draw a reprimand from the parks commissioner.

Major once caused a frantic stir. As I rode him across railroad tracks with my friend Betty leading him by the reins, he bolted at the cry of a train whistle. I was thrown off. Major easily pulled loose from Betty's grip and galloped off, leaving behind two frightened children with no hope of catching him. We had traveled a long way from home by then (taking turns riding or leading him) and thought we were near my aunt's Red Feather Farm in Bedford, which housed a candy business. But it was growing dark and there was no sign of her house. Not far ahead, however, lights were twinkling in a grocery store and the kindly owner came to our rescue. He sat us on a large sack of potatoes while he notified the police who in no time located our parents. The next morning my unforgiving older sister, Priscilla, found Major miles away, chewing a stranger's freshly cut lawn. Inseparable friends, Betty and I were not allowed to play together for a whole week.

I shall always remember our blacksmith, Mr. Terhune. Major was a favorite customer of his. He always greeted me warmly and found a spot for Major in his smoky shop next to the livery stable, no matter how many draft horses awaited their turn. I usually sat on a keg of nails near the door, watching the forge spring to life as pieces of iron turned fiery red before being slung onto the anvil and shaped skillfully into horseshoes. To this day I can hear the pounding of Mr. Terhune's hammer, the iron ring of the anvil, and the hiss of steam as each shoe was finally cooled in a nearby tub of water.

Major in Maine

A view of Sagadahoc Bay, mid-1930s.

Teenage years turned my attention to larger mounts, for training, exercising or just pleasure riding, but Major always remained part of the family and joined us each year at our summer home in Georgetown, Maine. More vivid than a photograph are memories of cantering him across the flats as the tide was ebbing or of the times we hitched him to the old pony cart and rode to picnics on the outer beaches. The cart always held simple but ample provisions for hungry mouths, as well as tin pails for berry picking and

Major caught by an incoming tide, Georgetown, Maine, mid-1920s.

a fishing pole or two. On one unusually long foray, the incoming tide at Sagadahoc Bay got ahead of us, filling the tidal flats and channels before we could return home. Undeterred, Major seemed to enjoy it all. He willingly pulled the floating cart across the bay, two small passengers aboard and one or two others wading or swimming along side.

Major especially loved it when a friend of mine and her pony, Jimmy, joined us on our summer vacations, lending companionship and fun to riding partnerships. During one such visit, I had the fanciful idea of driving Major and Jimmy as a team. George Smith, a local fisherman who was a family handyman, was dubious and suggested that a double hitch harness would be too hard for me, a mere girl. I was determined to try to get it together. George gave in and helped me gather some pieces of old harness. I managed to double hitch the ponies to the skeletal framework of an old farm wagon. I used large workhorse halters for the breeches (or holdbacks) behind the ponies' rumps, and from the breastplates I attached lengths of rope for tracers to pull the cart. Reins were no

Major and Jimmy double-hitched to the old farm wagon, early 1920s.

problem. It was not elegant, but my double hitch held together and worked for several days—until we got stuck in a mud hole far out on the flats. One of the wheels broke off, leaving us no choice but to desert the dilapidated conveyance to the fate of the tides, mount up and canter the ponies home without it.

I was always happy when my parents brought along our faithful canine friends, Rob Roy, Maple Sugar and Choo Choo from Lexington. I remember the latter two frisky pups pulling me around by their makeshift harnesses made of ropes and rags. My sisters were horrified—until Gramma Noyes came to my rescue by making more comfortable harnesses for handling the small team.

My double-hitch dog harness on Maple Sugar and Choo Choo at Georgetown, 1918.

Mother also ran a small summer camp for children at our place in Maine, named Camp Sagadahoc after the bay it faced located at the mouth of the Kennebec River. This was during the 1920s. It offered swimming, life saving, crafts, horseback riding, boat trips and sailing. Campers learned the vagaries of wind-shifts and sailing safety in the camp's fleet of four small boats, each with one sail. Usually one of my older sisters or I were available to pitch in as extra counselors, and we often brought ideas from our experiences at strictly run Girl Scout camps.

A favorite craft at Camp Sagadahoc was basket weaving which was taught by Penobscot Indians who spent their summers on Popham

Rob Roy, our Collie, overlooking Sagadahoc Bay, Georgetown, Maine, mid-1920s.

Beach at the head of the Kennebec River. Their visits were governed by the tides and we looked forward to the days when we could watch them paddle up the bay. One member of this tribe showed us where sweet grass grew along the marshes and edges of inland fields, and we learned where to gather fragrant bunches to clean and dry before the weaving. The mere sight of a basket evokes for me memories of these native craftsmen. They not only taught us with skill and patience but also shared stories about their forebears. From them we learned that Sagadahoc means "clam eaters." This particular bay had long been known for its abundance of choice white clams which could easily be dug in our end of the bay. The absence of any freshwater inlet into the central part of the bay made extracting them from the firm mud-sand flats easy.

A Dark Moment—and Guardian Angels

Even in our safe summer environs, it must have been hard for Mother to keep an eye on so many girls, but keep an eye she did. Once, one of the Indians who taught us basket weaving invited me on a canoe ride. Delighted, I readily accepted, and we paddled softly up the creek along the other side of the point. Suddenly, coming from behind, I could hear Mother loudly calling, "Come back, come back!" I was puzzled and irritated, for what could be more exciting to a 12-year-old girl from Lexington than to glide along a creek powered by the quick silent strokes of an Indian of my Downeast paradise? I couldn't understand why Mother was calling. Now, as a mother and grandmother, I do know why. Is it possible I knew even then? Perhaps I did but chose not to think about it—because of something that happened earlier in Lexington that I never told anyone about—not even Mother.

I mentioned previously taking Major to our blacksmith, Mr. Terhune. He was a kind Irish Catholic man, and Mother trusted him to keep an eye on me when I took Major to his shop for shoeing. This smithy was indeed a safe place, but the livery stable located next to it was not, not for a young girl alone. The men and boys who worked and passed time in it were of a rougher sort than I usually associated with, and they were not shy about paying unwanted attention to girls, as happened one day to me. My friends and I often stopped our ponies outside this livery stable to watch horses being groomed or harnessed. On this particular day, however, I was alone when I stopped. Only one older man, whom I had seen there before, was present. He was friendly toward me as usual, but that day he had an odd look in his eye as he grasped Major's bridle in one hand. He moved closer to me, and quickly, furtively, slipped his free hand inside my blouse and began to grope me. "Stop," I cried, "let me alone!" He did stop, quickly released his hold of the pony's bridle, and I headed for home at a gallop.

Some 20 years later, after the war, after I had married and settled into this old farm I so love, there was a knock on the kitchen door. I barely recognized him at first, but indeed it was Mr. Terhune, our old blacksmith, now quite elderly. Time had not changed the kind aspect of his weathered face. I invited him in and offered him tea.

"Rachel," he said, "I'm old now, and I just wanted to see you before I died. I have such wonderful memories of you as a little girl, you and your black pony, Major, and the kindness of your family. It's so nice to know that you have a family of your own and still love horses."

I could see why Mother trusted Mr. Terhune to keep an eye on her tomboy daughter. He was watching out for me, although I knew it not. How many others, I have begun to ask myself, have watched out for me without my knowing?

Here another story comes to mind about summers at Georgetown. Sagadahoc campers were occasionally taken by motor boat to the lighthouse on Seguin Island, five miles from the mainland. The lighthouse keepers were always glad to see us, share our lunches and hear news of the mainland. Teenagers were allowed to climb the nearly vertical stairs for a good look at the huge lantern with its thick polished lenses. There was a small platform around the tower and we hung onto its iron rails tightly.

One time I begged off from the trip to Seguin, my excuse being the tendency to get seasick. Popover, the larger pony under our care, needed shoeing. I persuaded Mother to let me ride him to the blacksmith on our side of the river across from Bath, Maine, instead of going on the motorboat trip. The 14-mile distance was not daunting. Popover was quite up to it and I knew him well.

Occasional passing cars greeted us warmly along the way and by noon we had reached our destination. The farrier gave the pony some water and half a quart of oats, and began without delay fitting him with four new hand-forged shoes. Meanwhile, I sat on a bench outside munching my crushed sandwich and watched the broad-beamed ferry crossing the river out of Bath. I could hear the soft crunch, almost a groan, against the pilings as it nosed in for docking and the release of passengers.

I paid the kind blacksmith 50 cents for each of the shoes and mounted the refreshed pony. With his ears pricked forward, we headed home. The tide was running high after we crossed the first bridge. We had a clear view of small boats in full sail along the river and it was good to canter comfortably along the dirt road. Here and there farmers and housewives waved from their gardens or hayfields. Then came a steep incline down to the second bridge, a rattling wooden span separating the islands of Arrowsic and Georgetown and a long climb up the other side. The old stone schoolhouse soon came into sight and just beyond was the ice pond. In winter, large blocks of ice were harvested there and packed in sawdust for warm weather use. That's when Popover, with no signal from me, quickly turned left and ran the final half mile to Sagadahoc Bay and a happy reunion with the other campers.

When I was almost home I saw a car I had seen out of the corner of my eye once or twice earlier that day. This time I got a closer look and smiled to recognize our close family friend Roger Greeley who was our guest in Georgetown at that time. As I would later learn, Mr. Greeley, at Mother's request, had shadowed me in his car and checked up on me all day as I journeyed to and from Bath. I've come to realize that my life has been full of such guardian angels.

A Warning Ignored

Among my most vivid Maine summer memories is the evening of September 2, 1924. In retrospect, it may have taught me to listen to myself no matter what others say. On our holidays in Georgetown, Father would usually spend the workweek at home and join us on weekends. He would arrive from Boston on the *City of Rockland*, a large passenger boat that offered easy transportation along the Maine coast. It served for years as transport for our family and associated animals. To get Major back to Lexington, he was walked two miles to Bay Point and sent to Bath aboard the *Virginia*, a smaller Kennebec River boat. In Bath he was transferred to the *City of Rockland*.

When Father departed on Sunday nights, he would watch from the deck as the boat sailed out of the mouth of the Kennebec River and into the ocean before turning south toward Boston. But on that particular night, a dense fog lay over the harbor as the *City of Rockland* left Bath with Father aboard. The fog was too thick for our usual fiery farewell—a bonfire on the beach. A short while later it felt unusual not to hear the loud whistle that signaled the ship's southward turn to Boston, but we assumed that the fog had delayed its departure.

I was in bed for the night at 8 pm and lay restlessly listening to the voices and laughter in the living room below. George Smith was holding court telling Georgetown stories. Drifting through the chatter came the lonely bellowing fog horn and whistle of Sequin

Light, answered by a continuously ringing bell from Pond Island. Suddenly, very faintly, I heard a new sound coming from the Kennebec River. Shorter whistles were interrupted by longer ones and then short ones again. I listened intently. Could this be an SOS, the Morse code I'd just learned in the Girl Scouts? I listened harder. • • • — — — • • • (dot-dot-dot/*dash-dash-dash*/dot-dot-dot). Yes, I was sure of it! I tiptoed down the staircase and leaned over the railing to interrupt the grownups.

"Mama," I called, "I think Papa's boat's in trouble. I just heard it calling for help. I'm sure it was SOS!"

George didn't like being interrupted, certainly not by children. (He was the one who doubted I could do a double hitch for the two ponies.) The grownups looked at each other and fell silent to test my report, but the wind must have shifted. All that could be heard were fog horns and whistles. George dismissed my warning as the pipedream of an 11-year-old girl and Mother shunted me off back to bed. I complied of course, but my feelings were quite hurt.

Early the next morning came word of the disaster. We raced up hill overlooking the Kennebec River to see what had happened. Below us the *City of Rockland* lay astride a reef on Parker's Head. There was a gaping hole in her bow and the stern was sinking. It was her final sail. Fortunately, neither Father nor any of the 183 passengers were injured. Indeed, my father's chief worry had been rescuing the cat he was taking back home to Lexington. Thank goodness Major was not scheduled to return to Boston on this particular voyage.

Today children are taught to listen to their own feelings and voices. It was just the opposite when I was a girl, and it wasn't easy to swim against the tide. I heard the distress call and inwardly I knew I was right even before the world confirmed it. I tucked that vivid experience away deep inside. For all I know it has quietly fortified me during difficult moments and choices in my life ever since.

My father survived the wreck of the City of Rockland, September, 1924.

Farewell to Major

Before I leave my memories of our Maine summers I must bid goodbye to the memorable Major. He remained with our family as he grew old. One summer in the mid 1920s, it was evident that he was getting stiff and did not like being used (or misused, as we feared) by our young visitors. Now, with the loss of sea transport after the sinking of the *City of Rockland*, our family realized that getting Major back to Lexington would be a longer and more difficult journey for him. Mother considered leaving him in the care of local residents over the winter, but we couldn't bear the thought of Major up there alone and possibly jerked around by careless children. We hated to put him down, but Mother wisely saw that we had no choice. Life, she believed, should have value, for animals as well as people. She asked George to end Major's life, as painlessly as possible. That meant shooting him. At moments, Mother needed resolve to go ahead, but she said she knew that she was doing the right thing in the most humane, caring way.

One day Major appeared outside the kitchen. He was led by George, holding a rifle. He asked Mother where and she pointed to the sandy beach. Major died munching his favorite biscuits and never felt a thing, Mother said. He was laid to rest among the dunes. This beloved black pony was many things to me. He was my friend, companion, confidant and occasional co-conspirator. He was an intermediary between me as a child and the world. He taught me life's joys as well as the responsibilities that all of us must bear. How could one small animal give so much to the love and meaning of life?

Chapter 2

LEAVING THE PADDOCK

My love of animals and horses continued, but other interests and activities blossomed during my early teens and in high school. I played different sports each season and was active in the Girl Scouts. By family tradition, this included a commitment to the drum and bugle corps. Older sisters, Priscilla and Fordham, were drummers and Deborah was a drum major. I played the bugle. State competitions kept us on our toes. Lest I forget, there was school—English, Latin, French, history, math (geometry or algebra or physics) and art. I nevertheless found time now and then to sculpt little figures of horses out of Ivory soap cakes, a hobby that I loved. My true interests often managed to find expression one way or another.

In those days I also belonged to a church youth program organized to provide social and spiritual enrichment and fun. It was the Young People's Religious Union (YPRU) of the American Unitarian Association. It sponsored conferences or retreats on Star Island, one of the Isles-of-Shoals off the New Hampshire coast. Those beautiful, remote islands seemed to spark romance. I experienced my first real crush there on a young man who ran a farm in the Midwest. I loved being around him but if he knew of my feelings, he never showed it!

Several romances which started there did come to fruition however. One was between my sister Deborah and the man she later married, Dana Greeley, an internationally known minister who in the 1960s was the first president of the newly named Unitarian Universalist Association. I met my own husband, Mark, long after my YPRU days. He was from the South, the son of a Baptist minister, so he had never set foot on Star Island. Mark heard so much about the island romances that he said it must have been like shooting fish in a barrel. He wondered how I had escaped.

Gathering Greens

As summers ended and camp was closed for the winter, family activities in Lexington turned to gathering greens for wreath-making. We made holiday wreaths for the seasonal shop the Girl Scouts ran, and took private orders on the side to earn a little extra money. Come December, our house was fragrant with the scent of evergreen, as branches of fir balsam, white pine, cedar, spruce and hemlock were spread over the dining room table. Giving color to each wreath were bright red alder berries, which we had cut earlier in Maine. They had been carefully rolled in damp newspaper to preserve their freshness.

We also made small partridge berry wreaths for the holiday table. Taking care to embed the roots of the small plants in spongy sphagnum moss, we hoped to keep their cheery message alive through cold winters. How well I remember gathering partridge berries in our favorite Lexington locales as a child and teenager. Unfortunately, it was not long before many of the shaded woodlands were wiped out by developers making the patches harder to find. The memory of our berry gathering is kept alive by a favorite verse that Father wrote. The reading of this is a family tradition today.

A Partridge Berry Wreath

Skyward the tall pine lifts its head
And drops its needles for our bed,
Where, glossy green and cheery red,
Our lowly leaves and fruit are spread.

Now in this wreath we come to you
As loving friends are wont to do,
Their Christmas greetings to renew –
Accept, please, our good wishes, too.
      —Hollis Webster

Riding Days

Horses, horses, horses. The loss of my dear Major hardly ended my love of them. A day rarely went by, according to my diary, when I wasn't astride a horse—either one of our own or someone else's that needed exercise or training. In daylight or dusk, rain or snow, it mattered not. I was in their company. Riding through woodland trails and open fields, with or without company, was just part of my life in those days. Local horse shows also kept me and my pony friends together in friendly competition.

One privilege I gained as a result of all this was that of riding behind the hounds as the "whipper-in" for the nearby Vinebrook Hunt Club. When the club hosted matches against rival clubs, I announced the start of the competition by playing "Call to the Post" on my bugle. I was usually astride an uneasy horse. I recall the hunt master as a quiet if not withdrawn man. As it happened, he was madly in love with my sister Fordham and proposed to her. Some time later, we heard that he was found lying face down in a sandpit, on the site of what today is the Burlington Shopping Mall. A gun was found. Some people said that the hunt master was murdered, but I thought the sad truth was that he took his own life because Fordie wouldn't marry him.

My brother Albert was not always supportive of my equine interests, no doubt because he served as water boy for my horses in our own barn. An English teacher once told me that Albert had earlier warned her that when his youngest sister finally came along, even her themes would smell horsey!

Horses were involved in the highlight of my high school years. On spring vacation in my senior year, my cousin Betty and I were treated by her father to five days at Longfellow's Wayside Inn in Sudbury, Massachusetts. We rented horses from a Lexington stable and rode 16 miles over dirt roads to the historic inn. We roomed on the less pricey third floor, supposedly its former slave quarters —shades, no doubt, of some inglorious past history. Our horses were stabled in the inn's barn at night and by day we explored every trail we could find in Sudbury and its environs.

Riding in front of the historic Wayside Inn, Sudbury, Massachusetts, 1930.

My horse was a handsome, sturdily built, black and white animal. On our longer rides, however, I found him less comfortable than Betty's horse which moved with more spring in his step so we often swapped mounts. I wondered what could account for this comfort difference. I didn't know then, but the question lodged in the back of my active mind from that point on. I now think this awareness helped spark my interest in how bone and joint structures influence the gait of dogs, launching my research, writing and lecturing career on the subject.

Jumping at a Lexington horse show, 1930.

Self Portrait as an Artist

It was my junior year when Mother and Father began thinking seriously about my education after high school. Family tradition steered me toward Radcliffe, from which Mother, Priscilla and Fordham had already graduated. My other sister Deborah was then a junior there and brother Albert was a student at Harvard. These siblings followed in a direct line going back to our ancestor and 1649 graduate of Harvard and its fifth president, the Reverend Dr. John Rogers.

Around the time my parents were addressing the question of my going to Radcliffe, our neighbor and devoted family friend, Roger Greeley, (my guardian angel in Maine), suggested that I might consider art school instead. Mr. Greeley was an architect who was well versed in the field of art. I knew him as Uncle Roger because of his closeness to our family. He became the father-in-law of my sister Deborah after she married his son, Dana. Roger was also my Sunday school teacher. One day in class I mixed up Muhammad (then typically spelled *Mohammed*) with Nebuchadnezzar, the builder of Babylon. From then on, Uncle Roger always called me "Mo."

Uncle Roger was interested in my soap carvings of horses and dogs and he knew that I loved drawing and enjoyed art classes. He also knew that I didn't shrink from a challenge. After the stock market crash of 1929, the state sponsored a competition for posters that illustrated the virtue of thrift. Uncle Roger encouraged me to enter the contest and I won an award. To my surprise he showed a few of my drawings and rough sculptures to a colleague and sculptor, Cyrus Dallin, the head of the Boston School of Art. He was the artist who created the sculpture of an Indian on horseback "Appeal to the Great Spirit," which stands in front of the Boston Museum of Fine Arts. Mr. Dallin said I'd make a fine candidate for art school.

As a result of Uncle Roger's friendly intervention much discussion ensued as to which path I would take—college or art school? Given the influence this decision could have on my life, my parents decided to seek the counsel of a third party, Father's cousin, Daniel Chester French. Father wrote him at his studio in New York City asking if he would be kind enough to gauge my potential as an artist. He said "yes" and plans were made for me to visit. Although I had seen him only infrequently, the famous sculptor was know as "Uncle Dan" within our family. We shared the high esteem all had for him, in particular for his works in Concord, the Minuteman statue at the Old North Bridge and his statue of a seated Ralph Waldo Emerson in the town library. One of my personal treasures to this day is a small, white, cast-marble sculpture entitled "Retribution" in which a hefty Newfoundland dog stands over a small terrier that apparently barked once too often. This piece earlier graced the bookshelves of my childhood home, a gift to Father from the hand that made it—that of his cousin.

Uncle Dan (famed sculptor Daniel Chester French), 1930.

Daniel Chester French undated sculpture of dogs called "Retribution."

The Decision

It was bitterly cold that early January morning in 1929 when Albert took the wheel for the long drive to New York City. It was so cold and windy on the ride that we were under blankets in our drafty, old Ford station wagon. We planned to stay with a younger cousin who lived in New York. This overnight would prove to have an educative value of a different kind. Our cousin was not kidding when he offered to give us a "taste of the town" that night. He led us down a dimly lit street, stopped at a shop with a "Closed" sign in the window and knocked quietly on the door. Through a peephole a soft voice asked for identification and then gave the OK to enter. We headed to a backroom where a small group was enjoying a bit of liquid refreshment that was at that time banned by Prohibition. I was in a "speak easy!" Recalling our entry, I understood for the first time how such places got their name.

I was a bit nervous by the time we arrived in the city and made our way to Uncle Dan's studio. We were greeted by a short, dignified, older gentleman with a kind face and a neat mustache. Touched that we had come such a distance to see him, we chatted amiably about his New Hampshire connection with our family and his many friends in Concord.

His high-ceilinged studio, which was filled with statues and studies, was airy and crossed by shafts of light and shadow that seemed to catch fine particles in the air. In the center was an unfinished but still formidable model of a large figure, so large it overshadowed the artist. On the wainscoting around the walls were mounted profiles, busts, anatomical studies of arms, legs and small images of horses and dogs. I recall gazing at a large hand resting on the knee of a trousered leg in a relaxed pose. It looked vaguely familiar and I soon recognized it to be a miniature replica of the famous sculpture of Abraham Lincoln now enshrined in the Lincoln Memorial in Washington, D.C. Here I was with my Ivory soap figures of horses and dogs!

Uncle Dan nevertheless took his time perusing the figurines I'd brought, most of which were horses. He seemed genuinely interested as he looked over each piece with his practiced eye. At one point, he smiled (as I did, for sheer relief), when he said he could tell that one appeared to be "a pony turned into a dog." Then, after what seemed an eternity, but was likely only seconds, he gave his honest and tactful opinion.

"My dear," he said, speaking slowly, "these works are very interesting. Each tells a story. I know that what you are doing is giving you a lot of pleasure. You may of course go on—and I hope that you will—until you realize your limitations."

The decision was over. His advice clinched it. I applied for admission to Radcliffe. The next morning, Albert, who had been guarding my innocence during our stay in New York, appeared greatly relieved when we got back into our old, airy Ford wagon and departed for Massachusetts. And so was I.

Chapter 3

EXPLORING NEW TERRITORIES

I did go to college as I resolved after my visit to New York—but not until after a lesson in facing life's setbacks. To my surprise and dismay, I did not get into Radcliffe after my last year of high school because I had done too poorly on my SAT tests. I didn't know why, since the areas tested were in my strong subjects. Whatever the reason, the result was an unanticipated delay in my higher education. This setback wounded my pride, especially given the Radcliffe tradition in my family. I soon recognized, however, that it was all for the best. Mother helped with her optimistic belief that, as she often put it, "There's good in everything if you look for it." She preached the corollary of this: that bad as things are, they could be worse. Waiting also helped my father plan for my college tuition, the fifth he would have to pay for, and this one as the Depression was taking its toll. Confirming Mother's wisdom, the delay opened up a wonderful personal opportunity—joining Priscilla and two friends on an unforgettable, three-month road trip across the American West in the summer of 1930.

Priscilla, whom I called "Pike" (short for her earlier nickname of "Pickles") was my oldest sister. She planned the trip with two girl-friends from her high school days who agreed to my last-minute addition to the group. As far as travel goes it was rudimentary with the four of us squeezed into an open Ford touring car and plans to stay at dollar-a-night "auto camps." What our trip lacked in luxury it more than made up in sweep and grandeur. For 11 weeks from June to September, we would drive by ourselves from one coast to the other and back again. We savored such storied locales as Independence Pass and Aspen, the Bryce, Zion and Grand Canyons with their breathtaking forms and colors, Hollywood and Beverly Hills, Yosemite, Lake Tahoe, Mount Hood and Crater Lake, Yellowstone, the Tetons, Sundance, Sioux Falls and the badlands of the Dakotas.

A small act of kindness before we left meant much to me. A friend of my father's gave me a Masonic button to wear in case we needed a helpful hand on our trip. I knew little about the Masons, but I wore the button faithfully. We often encountered kind strangers who offered help unsolicited, many of whom remarked on the button.

Our reliable conveyance was our sturdy Ford Model A which we called "Tigger." It held our five suitcases, a small gas stove, pots and pans, blankets, two folding cots, a small tent and a tool box for changing tires—the aggregate of which did not leave much room for passengers. Undaunted, Pike and I set out alone from Lexington on June 24, 1930. Our travel mates, Sarah Emily Brown (called "Say Say") and Millie Fischer, met us the next day in upstate New York at a farm owned by friends. The summer of 1930 may have been a discouraging time for our nation, but I was 17 years old and thrilled to be starting such a great adventure.

Having grown up in the environs of eastern Massachusetts and the coast of Maine, the boundaries of New England seemed endless that first day we headed west. I was reminded of a map I had seen in a Boston newspaper lampooning the attitudes of people in New

England. Entitled "A New Englander's Idea of the United States," it showed the state borders of Vermont, Massachusetts and Connecticut extending all the way to the Mississippi River!

"Tigger," our conveyance West, a 1929 Model T Ford. I took this photo of (L-R) Millie Fischer, my sister Pike, and Say Say Brown.

It didn't deter my excitement however, as we wound our way through Worcester and the Berkshires to North Adams, where Pike and I spent the first night. After hours of driving the next day, crossing Cherry Valley and losing our way two or three times, we arrived at the home of the Brooks, our family friends in Pearl Creek, New York. After dinner, I excused myself and headed to the barn to visit and talk with their beautifully kept horses. Then I walked through the orchard plucking and munching cherries as I went and across the fields to the creek for which the town is named. Our travel mates Say Say and Millie arrived the next day. After thanking our hosts, we set off for our first stop, Niagara Falls. It looked just like the pictures I'd seen in old *National Geographic* magazines. What I hadn't expected was the ear-splitting roar, the oceanic sprays of mist and the sheer power of those magnificent falls—especially breathtaking as we watched at night from the bridge with lights aglow around the wide rim of rushing water.

The next day we decided to push as far west as fast as we could. I drove about 100 miles which seemed to bolster everyone's confidence in my newly acquired skill at the wheel. We collapsed that night on the floor of a dollar-a-night cabin, replenishing our strength the next morning with eggs from Pearl Creek, strawberries and fresh cream from local farms and my mother's bread and cookies from Lexington.

Snapshots of a Trip

Throughout the trip I took scads of photographs and kept a diary to record my observations and thoughts. As much pleasure as these words and images give me, I must accept that the detailed story within a story that they tell goes beyond the scope of this memoir. I will have to give the full account—horse by horse, wonder by wonder, and blowout by blowout—in my next book, or addendum.

In this one, I can only lift out memorable images or moments to represent our Western trip. After endless cornfields we cheered with excitement at our first glimpse of the Rockies. By contrast was the fearful sign in Utah warning of no gas or water for 150 miles, and the searing, triple-digit heat of Nevada's Death Valley where the day's rising hot air has nowhere to escape.

An account of our trip west would not be complete without including the excursion that my sister, Pike, and I took down into the Grand Canyon. We made the fourteen-mile mule trek down the Kaibab Trail to spend the night at Phantom Ranch beside the churning Colorado River. I remember the fireside chat with our guide and how thrilled I was when he loaned me his hat and chaps. My correspondence with him following this adventure extended the enjoyment of the trip. I still have his last letter expressing his wish that our paths would cross again and that he would save a special horse for me.

My sister "Pike" (L) and I descend the Kaibab Trail into the Grand Canyon on mules.

Here I am with our Canyon guide, Charlie, wearing his hat and chaps.

Even more memorable was seeing the fabled white, red, yellow and vermilion pillars of rock in Bryce Canyon from the back of a horse. We quenched our thirst with cool, fresh mountain water pumped by a windmill. Riding horseback with Pike on the Pacific coastline between Carmel and Monterey was a special treat. At Crater Lake it was thrilling to edge out on a rock overhang watching sunset shadows as the sun disappeared behind Oregon's endless miles of mountains. Strange how I remember hot oatmeal at dawn with ranch hands in Montana—the best I ever tasted and the joy of pulling into a town just before the post office closed to find a letter from home. In Colorado, we found a herd of wild horses grazing by a sparkling, ice-cold mountain pool, framed by the Rockies and clouds lit up from below by the setting sun. We stopped. I walked toward them, camera in hand. An old white one, ears erect, the leader most likely, began trotting toward me. The others followed, broke into a canter and then turned toward a grove of aspens inhabited by a family of gypsies where a smoky fire and a covered wagon revealed their campsite.

Mount Moran, Grand Teton National Park, overlooking Jackson Lake.

Another moment I still remember was dodging wild weather in Wyoming. On the last leg of our trip, we headed across northern Idaho and Montana for the Grand Tetons. The ride was long, hot and dusty. Just as the tops of the Tetons finally came into view, the weather changed. Threatening storms hovered over Jackson Lake and thunder started rumbling. Pike, ever alert, took heed, pulled over and with tire patch, quickly mended a hole in Tigger's canvas top. Blankets were loosened from the spare compartments on the fenders and stuffed into the already crowded back seat just before the rain came down in buckets and torrents. The roads turned into rivers and gullies, belying the silvery clouds playing around the peaks of the mountains. Water danced heavily on Tigger's roof but inside it was dry and safe and we were just happy to be a band of Massachusetts gypsies snug in our temporary home.

There are memories of the people I met, sometimes only fleetingly, often around horses. In Santa Barbara, California, minor repairs to Tigger permitted time to visit El Paseo Ranch, where cowboys displayed Western skills in horsemanship, roping, barrel racing and all the riding acrobatics Easterners hear so much about. I was particularly taken by the Western tack, or horse regalia, on display including double hitch harnesses, saddles and bridles, some of them trimmed with silver buckles. As I admired all this ornamentation, I struck up a conversation with one of the cowboys who told me that he chased wild horses for a living—not the sort of occupation one heard about back home. He had "no use for civilization," as he put it. After conversing for a while, he cordially invited me to move out West, marry him, and chase wild horses! If four years of college were not ahead of me, I just might have taken him seriously.

Aspen and Other Ghost Towns

Here, in somewhat random and impressionist style, are a few of my recollections of the poor mountain towns we saw in the Rockies, the funny things that happened, a warm visit with friends, a "roundup" pageant in Oregon that fulfilled any Western fantasies we ever had, and, of course, the horses I met and rode along the way.

The mountain towns we saw after Tigger pulled us safely over Independence Pass in Colorado, towns hit hard by the Depression and the loss of mining jobs, left a vivid mark on me. Tumbledown houses or shacks with doors barred and windows broken were inhabited here and there by those who could not afford or couldn't bear to leave the mountains. One town, Fairplay, offered a memorable display of spirit. In the town center was a monument to "Prunes" a hard-working burro that helped miners throughout its 63 years. Fitted into that monument, proud locals told me, is a stone from every mine where Prunes worked. Gratitude to and appreciation of animals always affects me.

On the same leg of the trip we came through miles of aspen groves to another deserted mining town, Aspen. Signs of the Depression were everywhere. Two or three old loafers sat on the sidewalk smoking pipes. In front of them a saddled range horse slumped on three legs and above was a faded sign for Aspen's one and only restaurant, "Jake's." The men stared as we piled out of the car, four hungry girls garbed in overalls, pants, knickerbockers, bandanas and sailor's hats.

Jake himself, met us at the door in a stained shirt, sleeves rolled up. We entered a nearly empty dining room with a long counter, behind which was a cast-iron stove and two haunches of beef hanging on hooks. As soon as we were seated, we realized how famished we were. Beef was the only item on the menu.

"Ain't nothin' else," Jake said without apology as he got out forks, knives and dishes. "But this will be the best hot steak ya' ever flopped yer lip over." He was right. As the large sizzling slices filled the room with mouth-watering aroma, we found ourselves not caring about vegetables, dessert or coffee. As we gorged down our food, Jake leaned on his grimy arms and plied us with questions.

"Travelin' all alone, just you four girls?" he asked. "No man along, neither?" When I asked for milk, his reply advertised his sidelight as bootlegger. "Babies drink milk" he scoffed, "but I've got some of the best 'white mule' ya ever tasted…so where ya goin' from here anyway?" Needless to say, we felt that his curiosity could have killed more than a few cats. We got the check, paid up, piled into Tigger and drove out of town as fast as possible. Back then, I wouldn't have recommended this cow town to fellow travelers, but I've heard that Aspen has come up a notch or two since 1930!

East Meets West (and Other Things that Made us Laugh)

I can't forget as we left Aspen how we upset the orderly passage of 4,000 sheep we encountered along the narrow mountain road. Its steep shoulders left no room for us and the sheep to pass. The sight was irresistible so we pulled over to get our cameras out. When we stopped, so did the sheep, en masse, scattering on either side of the car. Furious, the herder sent collies barking after them. Then he turned toward us, swearing.

"You blasted dudes," he yelled, "Don't you know if you stop, the sheep will stop too? If you'da kept goin', they'da gone right past you!"

We called out meek apologies as we steered Tigger, as carefully as possible, through the wooly sea, knowing that if we had been locals and done this, we might not have been forgiven.

Yosemite, in the northern California interior, was another spectacular spot punctuated by a funny moment. We had arrived in time to see the Yosemite "fireworks"—fiery shards of lava-coal cascading from Glacier Point to the valley below. The next morning, we climbed a 3,000-foot trail to the ledge on Glacier Point where park workers used shovels to pitch the fireballs off the cliff in a spectacular display for park visitors. The guide then gave a spellbinding talk about the geological formation of the magnificent ranges and valleys around us. We were awestruck and found it hard to believe that millions of years ago where we sat at that very moment, the land had once lain under a vast shallow sea. Unfortunately, the spell was broken when I sat down on Millie's lunch!

The clash of East and West was always fascinating, and sometimes humorous. At Zion Canyon, Pike and I hired horses and a guide to explore its brilliant red, yellow and white sandstone cliffs and rock formations—the strange beauty of which was indeed of biblical dimension. We rode our cowponies up the narrow gorge, across creeks and quicksand, until our path ended between sheer rock walls. We returned on a low plateau where the horses enjoyed cantering along the edge of the creek and jumping over downed logs. Our cowboy guide was in a celebratory mood and he said he had good reason. It was Utah Day marking Utah's statehood. When I admitted I never heard of it, he was surprised. I asked him if he'd ever heard of Patriot's Day in Massachusetts, which falls on April 19. Nope, never had. We were even.

In the same vein, a few days later, after surviving the merciless heat of a passage across Death Valley, we treated ourselves to shampoos and haircuts at a barber shop. The lady hairdresser peppered us endlessly about touring the West without a man. When we said we weren't heading back until we got a real swim in the Pacific Ocean, she shivered. "It's terribly cold!" she exclaimed. We explained that we weren't concerned since we had always heard the Pacific was warmer than the Atlantic. "It couldn't be warmer!" she said. "The Atlantic is fresh water!" We were both wrong! We did find the Pacific to be colder than the Atlantic—and it was salt water of course. So much for New Englanders who think Vermont, Massachusetts and Connecticut extend to the Mississippi!

At Home with Friends

Various family friends, advised ahead of time of our plans, opened their homes and offered welcome breaks from the endless miles of dusty roads. One fabulous visit was spent with the Fryes in Redlands, California who lived amid orange groves near the Pacific Ocean. Alexis Frye had been the owner of Ginn Publishing Company in Boston. He was in love with my grandmother's sister who worked for the company for many years. Their unfulfilled romance ended in a lifelong friendship, which was likely the source of his warm welcome.

"Uncle Lex," as we called him, led us on one adventure after another including our first plane ride ever. I remember how, after taking off from the small airfield, Say Say and I clung together as the plane tilted and headed over the ocean miles beyond Catalina Island. We had an excursion in a glass bottom boat, gathered oranges in his grove and visited the historic Mission Inn, which testified to the art and influence of the early Spanish settlers. Uncle Lex had one more trick up his sleeve, telephoning ahead to assure that we'd receive a warm welcome at Kellogg's Arabian Horse Farm in Pomona, California, our first stop after leaving his home. The huge, castle-like stable was surrounded by gardens and fountains, but was no match for the elegance of the Arabians with their long flowing tails and manes, and their coats of gold, roan, dapple grey, bay or chestnut. Two young foals with pricked pointed ears greeted us at the fence. They nuzzled us with their sensitive little noses so distinctive of the breed.

The Lore of the Round-Up

At the Portland, Oregon post office in late August, we were delighted to get mail from home containing $25 apiece promising additional horseback rides through parks we were yet to visit. I happened to see a poster advertising the famous Pendleton Round-Up, an annual pageant of Western lore. We headed for it in high spirits following the highway that parallels the swift Columbia River to a campsite near Horse Tail Falls and arrived in Pendleton just in time for the big parade. What excitement! Marching bands and horses were everywhere. There were cowboys and cowgirls on horseback, Indians in their finest trappings, small children (sometimes three on a pony), lines of covered wagons, old fashioned coaches and rickety buckboards—one of which was driven by a 104-year-old Native American woman. We bought seats in the huge stadium for the show that afternoon and were not disappointed. Chuck-wagon races, bulldogging, calf roping, steer riding and bronco busting all filled the program. That evening, the pageant telling the history of the land was beautiful, but also tragic in a way for it showed the coming of the white man in the West. Even this brief glimpse of real cowboys and real Indians made us realize that we were seeing the beginning of the end for a wild and noble way of life.

Indian encampment at the annual Pendleton Round-Up, Pendelton, Oregon, 1930.

The next morning as I approached the show grounds, I found Indian wigwams arranged in a city-like formation, with countless horses and wagons in front. I showed my pass at the back entrance. For me it was the chance of a lifetime passing through the gate and up the chute to the big ring. Save for one photographer, there wasn't an outsider around. A high-spirited bronco mare called "Home Brew" was being saddled for a private rodeo performance. The brave animal fought hard, nearly knocking over the anchor pony and throwing three riders! Finally she gave up as people jeered from the sideline, "Hitch her to the milk wagon." She cantered across the field and I followed her out of the corral and thought to myself how could they belittle such a tough fighting horse? I wandered back to where hundreds of happier looking horses were hitched here and there.

Heaven on the Back of a Horse

I had many wonderful experiences on horses (as well as mules) crossing canyons, ranches and riding in the parks. One unforgettable ride in Yellowstone National Park in Wyoming still stands out.

We had been greeted by large geysers as we entered Yellowstone the day before. We nearly froze in bed that night but were kept entertained by bears nosing loudly through garbage cans—ours and everyone else's. Later that day, when Say Say and Millie went hiking with two German gentlemen they'd met and Pike was shopping for supper, I strolled through the campground. I suddenly caught a whiff of a barn and followed my nose to where a man was watering horses. During a cordial chat about the horses under his care, he described a 5-year-old sorrel I might like to ride to the watering trough. You bet I would! So I got on, no saddle, no bridle, just a halter and hitch rope. However, the horse wasn't thirsty. He began to trot, broke into a canter, dashed under two lodge pole pines and began to buck. I was excited to be on a bucking horse in Yellowstone where all the other mounts looked like workhorses. I somersaulted to the ground, grabbed the rope and led him to the barn where I told my new friend that I'd ride him if he would let me saddle him. No sooner said than done. I was away on this lively five-year-old riding along the valley floor without a guide and probably breaking all the rules. I didn't care. Yellowstone never looked more beautiful than from the back of that horse.

The spectacular Mammoth Hot Springs, Yellowstone National Park, 1930.

This essay about joining a horse round-up in the Grand Tetons is drawn from my journal for September 9, 1930.

WRANGLING

HEAVY DAWN MIST stretched across the foothills of the Grand Tetons separating the timbered slopes and barren peaks over us from the range below. A pale full moon clung to the jagged, snow-covered crest of Mt. Teewinot—at 12,300 feet one of the tallest Tetons—even as the rising sun broke through clouds.

I had been invited to help round up horses on the range and drive them back to the ranch with a few local ranchers and cowboys. We set out in the crisp air of a September morning, the stillness broken only by hollow splashing and the dull clicking of hooves against rocks as our mounts forded the creeks. At the far bank, tails dripping and legs shiny, the horses crowded together as they lurched from the creek and trotted slowly onto the range. Once my cowpony, Ginger, pushed his chestnut nose in the water and, with a flick of his head, playfully shook a cold spray over the horses in front of him.

Sagebrush tore at our chaps as we rode along, our horses dodging the holes of prairie dogs and jutting rocks. Ginger tossed his mane and snorted loudly at any excuse to shy. A chilly night in the small corral at the ranch was enough to make any horse caper the stiffness out of its legs. The mouse-colored bronco of a cowboy named Bill apparently felt the same, as he suddenly began ducking his head, humping his back and bucking.

"Watch him, he's a tough one!" came a warning from one of the ranchers.

Bill was sitting well as his horse began to twist and jump, stiffening his leg muscles and landing hard on all fours. The harder he tried to throw him, the more Bill relaxed into the challenge. It was exciting but also upsetting each time I heard Bill's leather quirt whip though the air and slash down on the spirit of that splendid animal. Apparently this was a daily routine for Bill, whose success in breaking ponies was useful at ranches.

A portion of the range had been allotted to me, the guest rider. I was an experienced rider but an absolute novice at this kind of thing. Loosening Ginger's reins, we galloped toward a herd grazing some distance off, the very horses we were supposed to bring in.

As we came up behind them, I was quite ready for our quarry to bolt away, as in one of Zane Grey's wild horse chases. Instead, a handsome pinto with black ears lifted his head to face us but remained undisturbed. A heavy buckskin horse also took a look, but he too, fell again to eating. Did I seem that harmless, I wondered?

"Hey there," I yelled in my most commanding voice. "Get along you ponies!" But the wind and wide range swallowed my voice. I reined Ginger in behind a battle-scarred grey and waved my hat furiously. The old grey lurched forward, startling his companions, and all headed off in different directions. Quick as a flash, Ginger turned – I had no say in the matter – circled the stragglers and forced them back to the group. Then, under more prodding and guidance from Ginger, all began trotting toward the ranch.

I was amazed at the agility of the little chestnut under me. At the slightest pressure on his neck, he wheeled to round up the ornery roan and the lagging black and white pinto. I stroked his hot shiny neck and touched his sides with my heels. "Come on Ginger, let's get 'em going."

Across the range, the cowboys, having already rounded up their horses, were heading home and we needed to catch up. Ginger eagerly responded, galloping back and forth to keep the herd together. Now and then he stopped short, pivoting to avoid sagebrush and or flying heels. Finally they all broke into a canter and joined the others.

"This pony certainly knows his business," I said breathlessly to Bill. Leaning over, he affectionately slapped Ginger's neck. "For a first attempt," he said, "you wrangled that bunch in fine, Miss."

"Thank you," I replied. "They'd have come sooner if Ginger had gone after them alone." A broad grin lighted Bill's face. No further words were needed. By now the moon had disappeared behind Mt. Teewinot and the early sun's shadows hung over the sage ahead of us.

Once more the morning stillness was broken by the splashing and clicking of hooves as the horses pawed and stumbled across the creek. Then came a scrambling up the bank, a trampling over well-packed earth, shouts from cowboys and a crowding of excited horses through the corral gate. Dismounting, I uncinched Ginger's saddle, slipped the bit from his mouth, and turned him loose. His work for the day was done.

All good things must end and so did our Western adventure. After Wyoming, the Dakota badlands were dusty and sun-baked, but interesting. I was grateful for a breathtaking sunset in the Black Hills of South Dakota, but sad as well, because reaching that point confirmed our departure from the magnificent West. It was a long way home after that, as had been the long way out, but the return seemed even longer because it was punctuated by more tire blowouts. Although five days of driving remained, I felt we were almost back in New England when we hit the paved roads in Illinois and were welcomed in Wheaton at the home of Millie's relatives. Not until we were inside their well-lit and orderly garage did we realize what rickety specimens the four of us (and Tigger) were. No apologies were needed, however. Our hosts were relieved just to see us in one piece and became closer friends than ever through the sharing of our stories. And so it was when Pike and I at last returned to Lexington.

My high school graduation photo.

Onward to Radcliffe

While it felt good to be home, my return to civilization in September 1930 offered no escape from the books. To bring up my test scores and rescue the family reputation I attended Manter Hall, the tutoring institution located in Harvard's shadow in Cambridge, at which Father taught. I was reassured to meet several pupils my age in similar circumstances enjoying the relative freedom from the pressure of classroom schedules, I became qualified for my first choice of college with no problem. The next spring that special piece of paper that had seemed so elusive a year earlier arrived in the mail—acceptance into Radcliffe!

When I entered college in the fall of 1931, class of 1935, the entire country was suffering. It was probably the worst of the Depression, the severity of which we saw more clearly in retrospect than at the time. Banks and businesses failed, and there were bread lines and suicides. Our family knew of these tragedies but for us the Depression was a fact of life and we dealt with it daily. It may have helped that the self-denial and discipline the circumstances required did not come as something new to our family. We were the quintessential New Englanders—frugal and not inclined to personal indulgences. It's a small thing in the context of the times, but, as an example, during my childhood, long before the Depression, Mother kept a "sacrifice bowl" in the middle of our dining room table. It was for sugar which was expensive then and thus a luxury. Rather than sprinkling it on our cereal, we put our shares of sugar into the bowl that could then be used for special treats or a dessert. I've never since used sugar on cereal.

Our family home was within commuting distance of Cambridge, but my parents insisted that I live on campus at least for the first year of college. They somehow raised the funds for this to happen, feeling strongly that living on campus, even for only one year, would be a valuable introduction to the overall college experience and the beginning of lifelong friendships. They were right.

With my Radcliffe friends. I'm in the middle back row, directly in line with the tree.

We were Radcliffe's largest freshman class yet and the older students welcomed us warmly. We soon had to choose from an array of extracurricular activities including clubs such as those for drama or international affairs, athletic activities and the Choral Society. The latter was a sustaining inspiration to me at Radcliffe especially around Christmas, when we sang Bach's *Mass in B Minor* at Symphony Hall in Boston under the direction of Russian born conductor and composer, Sergei Koussevitzky. I also joined the Radcliffe Christian Association where my introduction to the group was reading a poetic passage during a program that was intended to inspire prisoners at the Charles Street Jail in Boston. I wish I could recall the exact passage, but whatever it was it was not well chosen. It included a line about the bravery of "leaping to freedom." Given that the prisoners seated before the rostrum were truly a captive audience, with guards on hand to keep them that way, a wave of titter swept the room and I was mortified.

My choice in athletics was to join a small band of polo pony enthusiasts. I liked it and remember making at least one goal, but my enthusiasm dwindled as our inexperience in swinging mallets caused injury to the patient horses. At $1 per practice, it was an economic luxury so I switched to more affordable and safer sports of basketball and hockey. Money was scarce throughout college. All of us sought odd jobs to supplement student loans and help with tuition. We sometimes found bits of work through the Works Progress Administration (WPA) which was not confined to public works. Once, the WPA paid me 50 cents to go to Central Square in Cambridge, watch a movie and write a summary. A terrible movie, but I got a half-dollar to watch it—a lot of money at the time!

I was elected class secretary and enjoyed this honor greatly despite a few minor lapses. At our first official meeting, I forgot the secretary's report! Another time, at a special class luncheon, I forgot to call for the president of Radcliffe, Miss Comstock, who was scheduled to speak. Fortunately, someone else did remember, and I breathed more easily when she graciously opened her remarks by recalling how, for a number of years, my mother made a Christmas wreath for the front door of our hallowed Fay House. Mother loved Radcliffe and was proud that both she and her daughters graduated from there. Another awkward incident occurred when I was business manager of the freshman play, and I managed to spill the cashier's tray of ticket receipts, much of it small change, into the breathless hush of the second act. I never heard so many coins hit the floor!

We had fun at Radcliffe but we could not escape current events and dealt with serious social issues around us such as racism, prejudice and poverty. A brilliant Jewish classmate I recall fondly, Minnie Alpert, wrote a stirring essay about Boston in which she contrasted the city's glowing cultural opportunities and wealth on Beacon Hill with the poverty that shadowed the North End and other areas. It ended in a warning to "Boston, a city with her head in the clouds and her feet in the mire." Minnie was small of stature and wore thick glasses. Her themes in English class were intellectual and well written. She had a cheerful smile for everyone and she was a good friend of mine. To the envy of classmates, our English instructor read her writings aloud as examples to which we should aspire. For good reason. Under the pen name, K. B. Gilden, Minnie later wrote a popular novel, *Hurry Sundown*, about racial problems in the South. She named the hero of the story, Scott, after our English instructor whom she adored.

Minnie and I were among 14 students from Radcliffe and Harvard who traveled to Virginia by steamship for the 66th anniversary of the founding of Hampton University, the college created to educate African-Americans. The visit opened our eyes to a disturbing but vital dimension of American history. A delegation of black students welcomed us graciously and gave us a tour of the campus. We shared ideas with them and other students, sat in on some classes and enjoyed a program of interpretive dance. The capstone event was an evening ball at which the Radcliffe and Harvard students enjoyed the grace and rhythm of quiet dancing with Hampton students. For a short time, I corresponded with one of my Hampton dance partners who indicated that he hoped to visit us in Lexington. Mother didn't encourage the idea, even though, as she knew, two close friends of mine in high school and at Girl Scout camp were African-Americans. A black student from Spellman College in Atlanta also lived with us for several months while attending graduate courses at Radcliffe. In spite of good intentions, the visit of my Hampton friend never took place.

After living on campus my freshman year, I commuted for two years from home, sometimes with Father in a two-seater 1927 Ford with a rumble seat. I returned to live at Radcliffe as a senior, earning room and board as an advisor to freshmen in a house next to the campus. I had become a pretty good driver on our 1930 cross-country trip and could handle an occasional mishap. Once during my college years as I rounded a curve while driving alone near our house in Lexington, a wheel fell off the old Ford and rolled the full length of Lexington Green along Massachusetts Avenue. I left the car on the side of the road, ran after it, brought it back, and, with tools from under the seat, put the wheel back on and went on my way, feeling rather proud of myself.

A second car mishap, also minor, is memorable for a coincidence that had the peculiar aura of fate. It happened in January of my senior year, during "reading period," the study period before exams. As an authority on herbs, Mother was in demand as a speaker so whenever I could, I drove her to speaking engagements and helped with the slide projector. On this January afternoon, I took a break from studying to do just that. On the way home, the old Ford roadster got a flat tire right in front of Mt. Auburn Hospital in Cambridge. It was easy to fix but a frustrating loss of time. Returning to my freshman house, where I served as a resident adviser, all was well, and I studied until midnight. Ironically, later that same night I was suddenly hit with acute appendicitis and was rushed by ambulance to the scene of my flat tire that afternoon, only this time I did not remain outside the hospital but went straight to an operating room. The coincidence was not funny as I was hospitalized for two weeks and missed all of my mid-year exams. This meant making up each exam later on, in addition to the usual finals. Somehow I got through it all and graduated with my class in June 1935.

Helping me along the way was one of my father's favorite sayings, which I jotted down on the first page of a diary that I received for Christmas in 1928. I would have a chance to put it into practice when I initially failed to get into Radcliffe. "Don't worry when you stumble," he used to say. "A worm is the only thing that can't fall down."

Student Council at graduation 1935. I'm in the front row, second from right.

My Radcliffe graduation photo, 1935.

Chapter 4

REGIONS OF THE HEART

One autumn, two or three years ago, I was sweeping leaves off the granite millstones leading to the front door of my old house when a stranger walked into the yard.

"I'm looking for Pagey," he said, "a Pagey Elliott." He was relieved when I said it was me, explaining that he had to go to the town office of Carlisle, Massachusetts to see if anyone knew where I lived.

"My name is Ritz Houghton," he said, coming quickly to the point. "I'm the son of the man you were engaged to for two years back in the 1930s. My father talked about you so much I've always wanted to meet you."

I was dumbfounded. As he drew close, his deep blue eyes and honest face left no doubt as to his claim. I put aside my broom and invited him into our living room. Where does one begin? Emotions welled within me as my mind flew back to the summers between Radcliffe years when my relationship with Ritz's father, Herbert Houghton, was closing in on a possible life together. Sixty years later his son was seeking answers as to why it never happened.

Herb had attended Wesleyan College in Connecticut while I was at Radcliffe. We worked together during the summers as counselors at Camp Sloane, a large camp in Lakeville, Connecticut, run under the auspices of the Westchester County YMCA and Council of Religious Education. Herb was head of the boys' division. I was head of the girls' junior division, the riding program and was also camp bugler. Herb became a comfortable companion, honest and dependable with a great sense of humor, and his skill at the piano made him popular with everyone.

The Depression was starting to ease when we graduated from our respective colleges in 1935, but money was still scarce and we needed whatever work we could find. We both accepted jobs with the same organization that ran Camp Sloane, each of us in the field of camp promotion and counseling young peoples' groups. It's interesting to note that when I was being interviewed for the position in this conservative, YMCA-run organization, the administrator suggested that I keep quiet about my being a free thinking Unitarian! Our salaries were low, not always paid on time, and I often had only 15 cents in my pocket for food.

Fortunately, this job didn't last long. After six months, I left to accept a better paying position as field director for the Boston council of Girl Scouts and director of a small camp for disadvantaged children at Cedar Hill in Waltham, Massachusetts.

Having shared similar working conditions and other experiences, my friendship with Herb deepened. He loved his visits to our summer home in Georgetown and I looked forward to his letters. There was no romantic fooling around however, even though, frankly, there were times I might have welcomed such overtures. I loved Herb but I didn't know if I was in *love* with him. My diary discloses that *once* I believed I must have been, but only once. Herb, himself, seemed to shy away from overt signs of affection.

Setting a Wedding Date

Our relationship slowly changed, and in 1936 we admitted we had feelings for each other, but was it true love? Herb did not propose directly. He took me to dinner and gave me a family ring that we understood betokened his intentions. His reticence puzzled me. It may be that he simply assumed we would always be together. Were there other reasons? Hints are found in the love poems that he secretly wrote almost from the start of our relationship. When he sent them to me several years later I was truly astonished, because I had never known about these writings or his true hidden love which he revealed only just before we became engaged. In the poems I find hints of inner conflict, a sense that he longed for intimacy but feared entrapment or the burden of future responsibilities.

We set our wedding for June 1938, two years later. Herb's family seemed happy with the news and welcomed my visits to their home. Their house was small, so it was necessary to put a cot in the family real estate office for me to sleep. I felt very uneasy with this arrangement, but I could not tell Herb or anyone why—indeed I could barely admit the reason myself. Early in the morning, before anyone was up, Herb's father paid inappropriate visits to my room to chat about nothing. Sometimes he tried to sit on the edge of the cot. Nothing untoward happened, but it took quick thinking and evasions on my part to ensure that nothing did. My fear of these early morning visits became a nagging strain when I visited with Herb's family.

My parents trusted the judgment of my heart, although being old-fashioned I'm sure that they would have been happier if Herb had asked for my hand in person. Instead, he mailed them an allegorical poem suggesting indirectly that he planned to marry their daughter and would take good care of her. Priscilla, my ever-practical sister, wrote me a brief congratulatory note tempered with an escape clause: "Cheer up, the two-year wait will give you time to change your mind." She had changed her own mind about suitors twice!

A Fateful Dental Visit

Herb and I were scheduled to work at a large national Girl Scout camp in the Wasatch Mountains in Utah, the summer after our June 1938 wedding. My role would be to supervise trail riding. With my impending move away from home, a dental check-up seemed advisable. My longtime dentist had passed away, however, and I didn't look forward to seeing a new one. Having soft teeth, I had spent altogether too many hours in dental chairs. But, one day in February, I took courage and went to the Commonwealth Avenue office of a Dr. Mark Elliott, the young dentist in Boston to whom I had been referred by our faithful family orthodontist.

I knocked on the door, a little doubtful because I had not called ahead. A voice beckoned me to enter. Dr. Elliott was sitting in his own dental chair, holding a large floor plan of the Statler Hotel in Boston. As chairman of an upcoming dental association meeting, he was allotting rooms for programs and events. He greeted me courteously and listened with interest to my reason for coming. He examined my teeth and, certain that I needed attention, scheduled an appointment early the next week, to be followed by several more. Dr. Elliott's demeanor was nothing less than professional, but he told me many years later that he never forgot the blouse I wore that day—a white one with colorful prints depicting the coronation in England the year before. I saw no connection at the time, but it seems worth noting that the May 1937 coronation of George VI was necessary because of the abdication of his predecessor, Edward VII. Eleven months after becoming king, he had gone against everything planned for him and married the woman he loved, an American socialite, Wallis Simpson.

Something happened to me on that first visit. Sitting in the dental chair, I couldn't take my eyes off his. I didn't know what it was and certainly couldn't describe it even to myself, but as I drove home I felt something stirring inside. I think he felt it too. I looked forward to my next appointment and have never dreaded one since. When the feeling remained even after a month passed, I knew I could no longer wear the engagement ring Herb had given me.

Dr. Elliott noticed immediately. "You're not wearing your ring today," he said.

"No," I replied. "I've decided not to get married."

His answer was short, not overly sympathetic and conveyed no interest in why. "That's all right," he said. After our next appointment, he asked me out to dinner. I was no longer Miss Webster, I was Rachel and he was Mark.

Mark with his first car in front of Forsythe Dental in Boston, 1938.

A few weeks later, I was with three of my co-workers at our summer place in Georgetown on a rainy weekend. Mark, who had been on a fishing trip in the northern part of the state, stopped on his way home, as he had previously hinted he might. My friends sensed a spark in the air and, after being reassured that I would have a ride home, tactfully departed a bit early. Mark and I enjoyed a memorable hike despite the rain, roaming the ledges and outer beaches along the ocean. For the first time in my life, I knew I was in love. I didn't need my diary to tell me so. He was, too. Did we dare admit it? We did and later began our courtship, and then wonderful partnership.

A Painful Parting

First I had to resolve the matter of my engagement. There being no doubt that I'd been planning to marry the wrong man, it was only fair to face Herb in person. The weeks after I reached that decision were stressful despite the reassurance of my sister Fordham who stood by with thoughtful understanding. It was a difficult and unsettling situation but I did not doubt for a moment that I was doing the right thing. At Fordie's suggestion, I finally took the train to New York where Herb was then a student at Union Theological Seminary near Columbia University. We met for lunch and talked uneasily. After my no-doubt faltering explanation, he looked at me directly. "Pagey," he asked, "Is there another man?" With tears in my eyes I handed him back his ring. We parted in good faith exchanging only a handshake and wishes for good luck.

Herb was married two years later and after completing his studies at Union Theological he had a successful career as a minister. Curiously, two of the three other suitors I had around this time in my life also became ministers.

Some time later, Herb telephoned me to say that his wife, Mary, was pregnant. He then told me, to my horror, that when he told his family the good news at Thanksgiving with Mary seated next to him, Herb's father turned to her, smiled and exclaimed, "Congratulations Pagey!" The poor girl! How my heart went out to her when I heard that story. Until illness struck in the last years of his life, Herb kept in touch with greeting cards and occasional phone calls, the latter somewhat restrained perhaps, but always welcome.

Chapter 5

BEGINNING A LIFE WITH MARK

After breaking off with Herb, I went by myself to the large national Girl Scout camp in Utah to supervise trail riding. After everything that had happened, it was wonderful to be out West, as Gene Autry sang, to be "back in the saddle again." It was hard to be apart from my family, who had been so supportive that difficult spring, but it was an enormous relief to have been honest with Herb.

The three-week national encampment in the Watsatch Mountains, called Camp Cloud Rim, was attended by Girl Scouts chosen as delegates from across America. They participated in a cultural interchange and explored first-hand the Indian and Spanish influence in the West. My immediate job was to organize daytime and occasional overnight trail rides for the girls. In addition to those activities, we enjoyed crafts, hikes and refreshing swims in the ice-cold mountain lake.

I was not alone, though. To help pay my way, four teenage girls selected as camp delegates from New England were entrusted to my care. We drove out in my new Ford convertible stopping at a few national parks along the way and we quickly became a congenial group. They took turns behind the wheel and all pitched in with the packing and unpacking of sleeping bags, suitcases, tents, cooking and rain gear. The girls were good sports, showing initiative, courage and a great sense of humor.

I'm leading a trail ride at Camp Cloud Rim, Park City, Utah, 1934.

I was glad for their company. The letters I received from Mark at planned post office stops along the way also strengthened my growing feeling that he would remain in my life. The girls were a terrible tease over my anticipation of his letters, but Mark grew in their affection as well because he sent candy or "Bolster Bars" for them along with my letters. In the years that followed, during Mark's war service and the raising of our children, I lost track of these fine teenagers. They still remain in my memory with love and hopes for their happiness and well-being.

Getting to Know Mark

Back in Massachusetts that fall, Mark and I had fun boating on the Concord River and taking occasional horseback rides on wooded trails. It took some persuasion to get him on a horse, but he soon became comfortable in the saddle and even accepted the challenge of an occasional canter on a well-behaved mount—until he gained a few pounds and his riding pants no longer fit. Rather than replace the pants, he gave up riding.

We were followed on horseback by the German Shepherd that Mark gave me that autumn, my first close acquaintance with this breed. He had swapped one of his favorite guns for the pick of a friend's litter. Tedo, as we called him, responded quickly to obedience training and it wasn't long before I caught the obedience bug myself. In 1939, I joined the New England Dog Training Club and enjoyed showing Tedo who quickly progressed on the CDX (Companion Dog Excellent) level.

The only problem with Tedo was his terrifying fright of thunderstorms. Shortly after we were married, we left Tedo alone in our apartment for a short time and while we were away a bad thunderstorm struck. We returned to find two window sills chewed to pieces and the shaken dog greeting us with tail between his legs. Thunder shyness is not uncommon in the dog fraternity, but this was our first serious experience with it. We didn't scold Tedo, of course, but offered reassurance and forgiveness. Unfortunately we lost him suddenly and too soon in a fatal car accident. Another loss, another empty heart. Everyone shared our sorrow—except perhaps the postman whom Tedo considered a daily intruder.

It was reassuring to see Mark's love of animals. I recall when he took me to see the "little kiddies" he told me about in his office not long after we met. Given my feelings for him, I was curious, even a little apprehensive that they might be his own. As it turned out he meant the children from poor Boston neighborhoods who received dental care for ten cents a visit from him and others at the Forsythe Dental Infirmary for Children located in Boston's Fenway district. In addition to being a Forsythe dentist, Mark and Dr. Percy Howe, the clinic's director, conducted research on the effect of fluoride and proper nutrition on pediatric dental health. Mark also helped with the care of the research animals and often invited his small patients to come down and have a look at the guinea pigs and a monkey housed in the lab in the building's basement. I didn't dare tell him, but when he showed me around one day, I was relieved to learn that his "little kiddies" were not from a previous wife. Mark, by the way, was popular among the Forsythe dental staff and I could tell the kids loved their "Doc Elliott."

Mark proposed in December 1938. He first told his family in Montgomery, Alabama, where he grew up and then my own parents. I know they appreciated that my husband-to-be called on them directly to ask for their daughter's hand. We were married the following September, on the 9th, a warm bright day, at the Unitarian Church in Lexington. Reverend Dana Greeley, my brother-in-law, performed the ceremony, and I was proud to walk up the aisle on the arm of Father. I was the last of his four daughters and Father often called me his "blue-eyed boy"—an affectionate reference to the fact that, before I was born, he had not expected a brown-eyed girl. He wove this family joke into the lovely poem he wrote and read at my wedding:

The Message of Herbs in the Bride's Bouquet

And as you go, wherever you may roam
May rosemary remind you of the home
That's sheltered you, and of your parents care
To see you happy and contented there
To keep you safe from any threat of chance
And shield you even from an unkind glance
To give their love has been their pride and joy
For they're devoted to their "blue-eyed boy."
—Hollis Webster

Our wedding, September 9, 1939, in Lexington, Massachusetts. L-R: Irvin Middleton, bride and groom, Rosamond Greeley and Betty Clayton.

At the reception with Tedo, my German Shepherd, 1939.

After the ceremony guests walked to the reception at our home across the avenue. Our blue car, which we had dubbed "Blue Gentian" after the flower of the same hue, was in the driveway and behind it was Mark's small boat giving away our honeymoon intentions. Where else would an avid fisherman wish to take his bride other than a lake in northern Maine? Of course, we stopped at Georgetown on the way back.

We rented an apartment in Belmont after our honeymoon, a happy introduction to married life—although it was not always easy for Tedo, whose self-control was challenged by the Boston Terrier that lived on the floor below us. After we had been living in Belmont about a year, Father suffered a heart attack and became bedridden and was cared for at home by Mother. At her request, Mark and I moved into the old family home in Lexington and remained for about two years until shortly after our first child, Betty, was born in April 1942. The son of a Baptist minister from the Bible Belt, Mark was surprised at first by the scarcity of Baptists in Lexington, but he took quickly to the town, making friends, joining the Masons and the Rotary Club and marching with the Lexington Minutemen on Patriot's Day.

Mark (musket over his shoulder) marches with the Lexington Minutemen.

The War Years in Indiana

We were still with my family in Lexington when the awful news of Pearl Harbor came, and the war began. Mark joined the Army Air Force Dental Corps as an officer, was soon promoted to major and assigned to a base of the Army Troop Carrier Command in Indianapolis, Indiana, where he served as base dental surgeon for three years. Moving to the Midwest was an unexpected proposition for us, but there was no time to think about it and no reason to object—the war was changing life for almost everyone, separating them from loved ones who were overseas or scheduled to go. Like everyone, we did what we were told.

There wasn't enough housing for married officers at Stout Field, so we had to locate living quarters nearby, which wasn't easy. Mark, who had gone ahead of me, found a likely place on the outskirts of the city and put it on hold. He then flew home in order to drive us back to Indiana with our baby daughter and our new Golden Retriever puppy. He had given me the puppy, Goldwood Toby, the prior year, partly to fill the loss of Tedo who was killed by a car some months earlier.

We were soon packed up, the loaded trailer behind our car and ready to depart. But leaving Lexington, we knew not for how long, was difficult. Hardest was saying goodbye to Father, who had suffered a stroke and was confined to bed. Words were not easy and not only because of his physical condition. He reached across the pillows and squeezed my hand. It was a final gesture to his "blue-eyed boy," both of us knew, and we did not hold back our tears.

When we arrived in Indianapolis, our prospective landlady, Nettie Cook, sized us up and seemed agreeable—until she saw a dog sitting quietly in the car.

"No dog," she said emphatically. "We won't rent to anyone with a dog. He'll scratch the floor."

Betty, Mark and Toby.

After worried words were exchanged, I opened the car door. Wagging his tail, Toby greeted Nettie quietly. Obeying my commands, he sat down beside Betty, taking up his post as her guardian angel. There was no further hesitation. Nettie, who had already approved of Mark, now approved of the whole lot of us and handed him the keys.

We were lucky to find this home within driving distance of Stout Field and gladly agreed to the inconvenience of sharing the washing machine with the two tenants who lived in the garage. A flock of thirty laying hens fell to our care, and we occasionally sold eggs at 35 cents a dozen. The vegetable garden, which I tended with the help of Nettie's good natured husband, Charlie, meant ample fresh vegetables. I was less keen about the intrusive eye of Nettie's mother, one of the garage tenants, whose ample bosom was often sprinkled with ash from her pipe.

When Charlie was not home, Nettie milked their one cow. Our vigilant landlady called everyone "sister," including the cow. As she milked the patient animal, we could hear "Whoa, sister, whoa, sister," as she squeezed her udders. To keep flies away on hot or muggy days, she threw an old blanket over the cow's back and somehow got its legs into the four pant legs she had cut from her husband's old trousers, tying the improvised leggings to keep them up. I have a photo to prove it! Our daughter Betty loved the animals and was fascinated by Nettie's milking routine. She was also a keen observer, as I learned after the second Golden Retriever we had purchased had a litter of puppies. One day I found the poor mother, whose name was Chip, tied to a door knob in the dining room. Little Betty was sitting beside her, pulling at her nipples, murmuring, "Whoa, thithster, whoa thithster." Chip, incidentally, was quite agreeable to this attention. While Nettie's other renters became quite fond of the Goldens, Nettie grew more tolerant.

Nettie's cow dressed in pants to ward off flies, Indianapolis, Indiana, 1945.

Our Golden Dogs

When Nettie finally consented to our bringing a second dog into the house, Mark hitched a ride in a C-47 plane to purchase a bitch puppy from the Bushaway Kennels in Minnesota. The C-47 had gone to St. Paul to pick up a glider, in which Mark rode on the way back with Banty's Pluto of Bushaway on his lap. Chip, as we called her, was out of Field Champion Banty of Woodend, the first Golden of her sex to win this title. She was sired by Ch. Goldwood Pluto, quickly earned her CD title and became a wonderful, productive member of the family.

Like others of her breed, Chip was a great retriever and wasn't fussy about what she retrieved. Once she dragged home a neighbor's doormat inscribed WELCOME. Mark said that was probably why she took it! Another time she brought home a coverlet from a baby carriage which, of course, was not funny. Luckily, a fresh snowfall helped us follow the carriage's tracks and we returned it with profuse apologies. The baby had remained asleep through the theft, fortunately, and her parents were more amused than angry.

Both Chip and her accomplice, Goldwood Toby, tried in their own way to atone for these misdemeanors. One rainy night some friends came to dinner, leaving four rubber galoshes on the porch. As they got ready to go home, however, three were missing. There was great commotion and exclamations of disbelief. A search in the dark outside turned up nothing until the prime suspects, Toby and Chip, led us proudly back into the yard to the missing articles. The dogs' natural retrieving instincts turned our guests' annoyance into forgiveness. For me, it was a fun search that deepened my admiration for the tenacity of both dogs.

Featherquest Goldens over the Years

It may seem hard to believe today, but Golden Retrievers, who were developed in England, were almost unknown in North America before 1930. This noble breed won quick acceptance for its companionability, loyalty and skill in retrieving water fowl and upland game. Indeed, it was Mark's love of duck hunting that led him to take a chance on a Golden Retriever puppy he heard about at Hank Christian's Goldwood Kennel in White Bear Lake, Minnesota.

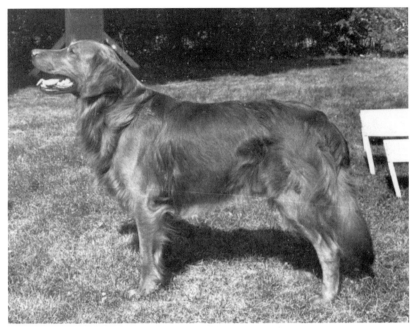

Goldwood Toby, canine patriarch of Featherquest Kennel.

I'll never forget the sight of Toby, this rich, brown, eight-week-old puppy on the front seat of his car as Mark drove into the yard in Lexington one day in 1941. This was Mark's way of trying to fill the emptiness left by the loss of Tedo. It was a complete surprise for me and we never regretted Mark's choice. Toby did more than fill a void. In 1945, he, along with Chip, laid the foundation for our own kennel, Featherquest. Goldwood Toby became the springboard of my half-century of adventures in the world of dogs—training, breeding, research, lecturing and, of course, enjoying the companionship of wonderful dogs and making countless friends.

Over the years I witnessed changes in type as Goldens grew in popularity. From darker, leaner working field dogs to lighter, more compact show dogs. I need only to look over photographs of my own favorite Goldens to see the variance in my own kennel.

GOLDEN RETRIEVERS

Early Featherquest publicity from 1946.

Goldwood Toby was an apt pupil in obedience training as a puppy and quickly showed promise for competition. He entered his first dog show in Massachusetts shortly before we moved to Indiana, winning best of breed over prominent local entries. Spurred by Toby's promise, I soon joined the Indianapolis Obedience Training Club and met other dog enthusiasts—a welcome social link during our somewhat isolating years there. As the first Golden in Indianapolis, Toby stirred considerable interest. That was no trouble for this fine dog since he was full of showmanship, loved to play to the gallery and made the most of each triumph.

Civic groups began to ask me to give obedience demonstrations and I remember one in particular that we gave to the local chapter of the American Association of University Women. Toby promptly did all I asked of him—coming when called, heeling by my side without a leash, sitting and lying down. I even had him retrieve a raw egg without breaking it. To further show the gentleness of his mouth, he picked up a live duck at my direction and carried it without ruffling the feathers. Toby also vocalized on command (but without barking) once, twice or three times—not an easy thing to teach a quiet dog.

His best "parlor trick" was fetching the letters of his name from 26 alphabet blocks strewn randomly on the floor. This was not a game of canine Scrabble, but to demonstrate his acute scent discrimination. One person would put down the letter blocks from which the four containing the letters of his name would be brought to me. I would then rub my hands on them, and my helper would toss them at random in the pile on the floor. At the command, "go find," Toby would sniff among the letter blocks and, invariably, bring me the four letters of his name, one at a time. His spelling was not perfect, but the audience was astounded.

With Toby, 1944.

Life with Nettlesome Nettie

Life under the critical eye of Nettie Cook wasn't easy. She found fault with my housekeeping, parenting and role as wife. A fervent evangelical Christian, she was intolerant of anyone not sharing her rigid beliefs. There was simply no use trying to explain why I didn't go to the weekly prayer meetings she held in her backyard in a converted henhouse that Nettie rechristened as a "tabernacle." While she never explicitly condemned my liberal religious beliefs (at least not to my face), she implied once that I helped provoke the Lord's vengeance when a tornado hit Indianapolis in 1944. Fierce winds tossed the swing off the porch and ripped branches off trees. It rained so hard that rivulets swished under the front door into the living room. Panicked, I looked for something to absorb the deluge, but any cotton rags or towels we had were already in use. Our second daughter, Ruth, was then a few months old, and a pile of clean, unfolded diapers happened to be in the living room. I grabbed some diapers and crammed them under the leaky door. Nettie arrived as I was cleaning up after the storm was over. She was furious to see diapers under the front door, although I was never quite sure why. In any case, she implicitly blamed us, muttering about God's anger at unbelievers and suggesting that we were somehow responsible for the tornado!

To give the devil his due, I must say that over time Nettie became tolerant at least of our dogs, to the point of not complaining about our first litter in the basement. Perhaps she recognized that the renters who also used the basement washing machine enjoyed watching the puppies play.

I had other cultural mishaps beyond Nettie's reaction to the storm. As a serious dog enthusiast, I correctly called female pups "bitches," but it wasn't until I visited the hairdresser one day that I realized such language could be misunderstood. I introduced myself to the woman who did my hair as a military wife with an interest in dogs.

As it happens, a close friend of hers was a local telephone operator, who could listen to any conversation over the wires. "Oh," the hairdresser exclaimed to my embarrassment, "You're the one who talks about bitches on the phone!"

Betty playing with the first Featherquest litter, Indianapolis, 1945.

Despite such experiences, our dogs and puppy litters did stir interest among friends, including those Mark and I knew at Stout Field. One of them was the base's cook, who now and then sent scraps from the Army kitchen home with Mark as treats for Toby and Chip. Truth be told, due to rationing, the scraps were usually more edible than any meat I could buy at a local market with my food coupons, and we never hesitated to swap the scraps for what we had on hand for our own table.

Finding River Road Farm

As World War II was ending, Mark expected to be sent overseas and had his bag packed, but the order never came down. He was transferred to an Army Air Force base in North Carolina to await discharge. It wasn't easy leaving friends we had made in our three years in Indianapolis, but it was exciting that world peace had come at last and everyone had a chance to get on with life.

Temporary living quarters at the family home in Lexington were not available, as my sister Priscilla was ensconced with her family of six. Luckily for us, Fordham ran up the family welcome flag in Swarthmore, Pennsylvania, where she lived with her husband, Dr. Alfred Calhoun, and their four children. Mark was being discharged from the service in South Carolina so Betty, Ruth (born in Indianapolis in 1945), and I gratefully moved in to the third floor of their spacious home. The cousins got along happily together, remaining cheerful even during an outbreak of chicken pox.

We had sent Chip to a training kennel, but Goldwood Toby was welcomed and the Calhouns' dog, Blackie, adjusted to his new companion. Fordie's wonderful housekeeper, Ethel, embraced the four of us, for which I was so very grateful. She and Fordie even insisted on my taking breaks from household activities to enter Toby in dog shows and obedience activities. The cousins enjoyed Toby's parlor tricks and I was later honored, at least I think I was, when the Calhoun children acquired a Basset Hound and named it "Pagey."

As our stay in Swarthmore stretched to nearly three months, we recognized how courageous the Calhouns' invitation had been. Eventually Mark got his discharge and we packed up and bid teary goodbyes to our generous hosts. We drove to Boston to accept yet another offer of family hospitality—this one a much shorter stay while we put our lives back in order, with Deborah and Dana Greeley at their home in Boston overlooking the Charles River.

Mark tackled the daunting task of pulling together his dental practice and I immediately began searching for a home for us. I'd always held warm thoughts about Carlisle, it being one of our family partridge berry haunts, but every real estate agent I phoned had nothing to offer there. A bit discouraged, I invited Mother's sister, Edith, who knew the area well, to help me search.

Aunt Edith suggested a Carlisle real estate agent named Sterling Davis, who was glad to be our guide. "No plain houses around here for sale," he told us, "just two or three old farms." We set out, and he took his time showing us two of the available properties, but neither appealed to me. By then it was late in the afternoon, I was tired, and Betty, who'd been in the back seat with Toby all day, was hungry and fussy.

"Well," he said, "if you don't like those two, you probably won't take to the other place I had in mind, an old dairy farm." I agreed and was ready to head for home when Aunt Edith turned to me and said, "Pagey, you're here, why don't you at least take a look." I thought that she made sense and agreed to see one more farm.

Overshadowing the house in question were two huge barns and two silos. Double-faced stone walls bordered the front and two sides of the house and stretched along the road behind it, telling of two centuries of hard labor with oxen and sledges clearing the fields. Pastures which had not been grazed for nearly three years spread toward the Concord River.

Brush overgrew the front door, temporarily barring entry. Once inside, I noticed slender twigs squeezing through the window jams. It was not perhaps what I had imagined, but it was true and beautiful. In my heart, I could hear the old dairy farm speak. Give me a try, it whispered to me, just give me a try. I turned to Mr. Davis and said, "Would you be willing to hold this until I bring my husband to look at it after supper?"

The broker was most willing to comply. That evening Mark and I stood on the side porch overlooking the fields, the serene beauty of the farm beckoned us. Ever the hunter, Mark noticed and pointed to a red tailed hawk circling the woods overhead. We both took this as a good omen.

Mark gave Mr. Davis a down payment of $100 and asked him to meet us the following morning with purchase papers. At $15,000, it was a little more than we wanted to pay, but we knew we could get help from a loan through the G.I. Bill and we signed. Only then did Mark and I learn that we weren't buying just the 10-acre corner lot. "Oh no," said Mr. Davis, "the property includes 64 acres, all the way down to the river." Mark and I exchanged disbelieving glances, squeezed hands and signed the papers.

River Road Farm was ours. Thank God for Aunt Edith! Had she not told me to take a look while I was there, we might not have bought it. But we did, and this magical old landscape became the foundation and terra firma of my life from that point on.

The farm is at the bottom of this photo. Our land extends to the Concord River.

Views of River Road Farm, summer and winter.

Over the years on the Farm there were times when we felt "land rich and cash poor." The sacrifices were made easier by our love of the land. The following essay describes perfectly how my family, and probably many others, feels about owning land.

I am real estate

I am the basis of all wealth, the heritage of the wise, the thrifty and prudent. I am the poor man's joy and comfort, the rich man's prize, the right hand of capital, the silent partner of many thousands of successful men. I am the solace of the widow, the comfort of old age, the cornerstone of security against misfortune and want. I am handed down to children, through generations, as a thing of the greatest worth. I am the choicest fruit of toil. Credit respects me, yet I am humble. I stand before every man bidding him know me for what I am and possess me. I grow and increase in value through countless days. Though I seem dormant, my worth increases never failing, never ceasing; time is my aid and population heaps up my gain. Fire and the elements I defy, for they cannot destroy me. My possessors learn to believe in me; invariably they become envied. While all things wither and decay, I survive. The centuries find me younger, increasing in strength. The thriftless speak ill of me. The charlatans of finance attack me. I am trustworthy. I am sound. Unfailing I triumph and detractors are disapproved. Minerals and oils come from me. I am producer of food, the basis for ships and factories, the foundation of banks. Yet I am so common that thousands unthinking and unknowingly, pass me by. I am real estate.

—Anonymous

Chapter 6

EARLY DAYS AT RIVER ROAD FARM

Some friends who lived in the Boston suburbs and were accustomed to manicured lawns and easy access to the city, could not fathom our choice of a rundown farm so far out in the country. But their children, who loved every visit to River Road Farm, thought otherwise. Soon we had ponies for riding and a pond for swimming and fishing. We also held informal dog breed matches, horse events, and even sled-dog races, and the open green fields invited everyone to walk and let their dogs run free.

After we moved in, the work we faced fixing the place up was daunting, but my mother was very supportive, cheerily reminding us "nothing ventured, nothing gained." She gave us much-needed courage to meet the challenge—and a challenge it was! With her enthusiasm and my brother Hollis's willingness and strength, we cleared the overgrown shrubs encircling the house and the noxious weed, polygonum, that almost hid the front walls. Mother warned us that the polygonum would take over in years to come if we did not get rid of it. Indeed, fifty years later, I see bits of this weed struggling for attention.

"Brother," as I often called Hollis, unearthed two flat grinding stones near one of the wells, pulled them around the house and set each in place as part of the stone entrance to the front door. He struggled with a mental deficiency caused by a lack of oxygen at birth, and he lived away from home much of the time. Brother was devoted to Mark and would do anything in the world to please him. Mark, who returned the affection, understood his affliction and was always aware of problems, or potential ones, that would require special care.

Hollis was different, of course. He couldn't follow the paths the rest of us took but he was a dear, and we loved him for who he was. Mother cared faithfully for him his whole life. Indeed her biggest fear, which didn't come to pass, was that she would die before him and he would be without her support. My own feelings about Hollis were loving and loyal, and I did my best to respect his privacy despite my uncertainty about what others might think. Having a handicapped sibling was poorly understood then. One thing about Brother was that he could surprise you. When we were adjusting to our new life at River Road Farm, he lived with us briefly. Once, as I was looking for a book in the living room, I noticed that several volumes were missing from the bookcase. It seemed to me that they were essays about religion or the meaning and power of faith. Several days later, I found the books. They lay on Brother's bedside table in his room. What interest he had in them, I did not know, but I remember thinking that whole worlds probably existed inside of him that even we who loved him would never understand.

New Life at an Old Farm

With the fields deeply overgrown from lack of grazing, we gladly consented when Dick Bates, a popular local farmer, asked to pasture his dry cattle in the well-fenced fields. The River Road dairy business had come to a halt both because of the war and the untimely death of the owner, Mason Garfield, a grandson of President James

Garfield. His pride and joy, the farm's purebred Jersey cattle had produced quality milk and cream for market. Mason knew every cow individually, evidenced by the names written over each of the 32 feeding stanchions in the big milk barn.

River Road Farm, late 1940s.

Those iron stanchions were removed to make room for pony stalls and hay storage, in the first of many changes we made to adapt the farm to our interests in breeding and training horses and dogs. We did this with great respect for the old farm's venerable past. The land was cleared by the first English settlers to move inland from Boston and they occupied this corner of the town they established and called Concord. The area later became part of Carlisle. The oldest part of the foundation of our house was built in 1701 for grain storage. In due time, change the farm we did. The large remaining barns, with bull pens below, were dismantled for second-hand lumber, and the two grain silos were hauled to a farm in a nearby town. Neighbors wished us well, but I know it must have been hard for some residents to witness such dramatic changes in their proud small town.

As a dedicated amateur radio enthusiast, Mark lost no time erecting a two-story radio tower above the dairy building. Connected by cables through a small hole he cut in the back of the house, his shortwave radio receivers, transmitting equipment, microphones and other shortwave equipment took over half the guest room. In the postwar years, when fears about the "cold war" with Communism were high and civil defense was an important responsibility for ham operators, my husband—using his call letters "W1YYI,"—was ready to help in any emergency.

How Mark loved those early morning hours he spent talking to friends near and far who shared his hobby. The hot coffee I would carry up the back stairs to him those mornings was always gratefully received, and now and then he put me on the air to communicate with acquaintances we had met during overseas travels.

Not long after he took over the guest room with all his radio equipment, Mark initiated a gathering of shortwave radio friends for the annual International Field Day, a worldwide competitive event for the most individual contacts within a specified 24 hour period. Preparations for Field Day involved setting up temporary headquarters in tents on the high ground overlooking the Concord River on the previous day with essential equipment and antennas attached to overhead tree limbs. This elevated location on the wide open field along the edge of the river was as far away from houses as possible. Resident cows, still in the pasture, watched the activities from a distance with quiet interest—interest that later turned to disturbing curiosity. One year when the ham operators returned early the next morning, they found two tents knocked over and some of the equipment upset. Nearby, the cows contentedly chewed their cuds showing no sign of guilt. All was soon rectified, but the chances of winning top place in the 24 hour time limit this year was no longer possible. Spirits brightened, however, as occasional chuckles were heard from overseas operators who learned of the bovine interference.

Enthusiastic hunting companions on land or water—Mark David and "Dennis" resting.

A happy addition to the family was our third child, Mark David, Jr., who in due time shared his father's enthusiasm for fishing and hunting. Much as he hesitates to admit it, Dave and his school chum, Dave Bott, enjoyed the ponies that were always part of the farm, particularly old "Brownie" who came to us with a four-wheeled back-to-back seat buggy. Brownie was as quiet to drive as he was to ride and the boys took pride in trotting the pony by themselves along the road. Dave's small cowboy boots worn when he was astride another pony called "Shaggy" still hang in his old bedroom. His favorite hunting companion was "Dennis," one of our early Golden Retrievers, who as a puppy chewed up David's favorite book, *Dennis the Menace*, and was promptly given the nickname. To jump ahead, on graduation from college, David became a pilot in the Air Force during the Vietnam War. As a second lieutenant, he was assigned to a unit of the Air Force Special Operations, known as the Ravens, their assignment carrying on the clandestine air war in Laos against the North Vietnamese infiltration. David was adopted by a Laotian family whose son he had saved from a serious infection—and was given their 12-year-old daughter in appreciation. The offer was graciously declined.

A Menagerie of Animals

Animals, big and little, were always part of our lives. One midwinter day, Betty excitedly brought a surprise into our living room: 24 chirping Rhode Island Red chickens, a gift from a local farmer. We kept them in the barn until spring, when our flock of laying hens was severely reduced by the discovery that half of them were roosters. The hens nested comfortably in Mother's old bee hive boxes stored in the root cellar under the garage. For two or three years we enjoyed enough fresh eggs at least for the family—until raccoons discovered these delicacies, scared the hens and chased them off. Having created an empty hen house, the raccoons showed good practical sense by promptly moving in with their four kits. The children were excited about keeping the young family, so Mark built an off-the-ground wire cage, leaving responsibility for their care to the girls. All too soon the kits were no longer babies and were wisely released a good distance from the farm.

Next, enter the goats. We had been persuaded that goats provide naturally homogenized milk that is highly beneficial for puppies. With two litters of pups on the near horizon, we decided to try providing this nutrient on the farm. A friend of ours, Sena Water-house, wife of Dr. Edward Waterhouse, Director of Perkins School for the Blind, persuaded me to take on the project. At the time Mark was a visiting dentist at Perkins. The Waterhouse family and Sena herself, who was blind, drank only naturally homogenized milk. Since it was not readily available on the market, Sena raised and milked her own goats. A remarkable woman! She knew each one by feel or voice, guided always in her movement by the German Shepherd that never left her side. On the strength of her advice, three goats soon joined us here at the farm. I learned how to milk them and our Golden Retriever puppies flourished.

Goats provided naturally homogenized milk for puppies, but also ate every-thing within reach.

Our only problem with the goats was fencing. They felt no respect for our boundaries, nor were they scared of the electric wire strung around the edge of their yard. One day, just to give a lesson in the consequence of trespass, I pushed one member of the flock against the hot wire. But the goat didn't get the jolt; she was only the transmitter who passed it to me! After that we let them have the run of the yard, with the barn door open. They ate everything in sight, including the Christmas wreath I had hung on the front door.

One day in late spring what looked like an apparition appeared in the yard—a large horned Billy goat, apparently looking for the ladies. I had no idea where he'd come from until a short while later the animal's relieved owner, who lived on the other side of town, drove in with his trailer. We'd had enough surprises, and after a while found a home for our three goats at an established goat farm several towns away.

The Trail You Leave in Ink

In 1964 I signed up for a correspondence course in a subject that had always stirred my curiosity—handwriting analysis, better known as graphology. Was it introspection on my part, or seeking a better understanding of the many friends who had been so generous in lending a hand along the way? Perhaps it was both.

Graphology is an art and science that deals with the space, form and movement of individual strokes in a person's handwriting that offer insight into the character and the personality of the writer. The study is dedicated to self knowledge and the understanding of others, their talents and their tensions. Studying handwriting gave me an appreciation for my own aptitudes and made me aware of the great gifts in the wide variety of people who were in my life.

There are skeptics, even today, but my confidence was unshaken and I was stirred into further study by a collection of books on the subject. To my surprise I learned that the history of handwriting evaluation went as far back as Aristotle in 300 B.C. and the Roman Emperor Nero who was said to have pointed to a man in a court hearing with the words, "His hand writing shows him to be treacherous." Little did I know how this course of study would impact my second daughter, Ruth. After living overseas for eleven years in West Africa and Belgium, she returned to the United States to raise her two children with her husband Peter, a Goodyear Tire and Rubber account executive. When she delved into my handwriting library in 1979, she became completely engrossed in handwriting and document examination to the point where she established Pentec Incorporated, a forensic and employment consulting firm that advises individual legal and corporate clients in the U.S., Canada and Mexico. Her daughter, Sarah, has since joined her company now with offices in Michigan and Massachusetts. Her son, Nick, is an art lawyer in New York City.

The handwriting correspondence course that I undertook was based on the details and meaning of forms, strokes, structure and movement shown in a person's unique writing style. The information seemed to add another dimension to the study I was already involved in related to the structure and movement in a dog's gait. Both studies revealed patterns—one was brain writing, the other structural and physical! After two years of study, I achieved certification as a handwriting analyst and, stirred by the interest of friends, I ventured into teaching an eight-week course here at the farm. I continued my learning through meetings of the local handwriting chapter and occasionally gave talks to outside groups.

Through the study I became amazed at how many strokes or formations have carried through the years from my own writings as a ten-year-old to this day—despite the autocracy of the old Palmer handwriting method dedicated to uniform letter formations for school children. When a member of our handwriting group, Lou Thackery, a well-loved professional handwriting analyst, took the time for a lengthy analysis of my cursive art, he asked for samples of my writing from years past. I found several dating from 1921 to 1999! The details of his findings were amazing, very frank, and as far as my future was concerned, showed aptitudes right on the mark—research, writing and lecturing.

I only wish I'd found time to continue the study, for the warning from other analysts was true: "If you don't use it, you lose it." Ruth and Sarah have done their best to keep me on track, but I admit, though memory may fail, the interest is still there.

Featherquest Kennel

The years with our three growing children were both busy and happy, sharing their school activities, riding, fishing, blissful summers in Maine, including David's enthusiasm for surfboarding on the open coastline. Along the way, my interest in dogs became more involved. Our first Featherquest litter born here at the farm stirred particular interest because of Goldwood Toby's background. Raised in the Goldwood Kennel in Minnesota, he came to us as a puppy when we were living in Lexington. He was sired by Ch. Toby of Willow Loch, the first American-bred Golden to win Best in Show in the United States, and descended from strong working Speedwell lines imported into Canada in the late 1920s. Toby was darker in color and a little different in type from some of the Goldens in our area, more recently imported from England.

Toby with first Featherquest litter in Carlisle, 1947.

With Ch. Featherquest Morning Sunray, 1970.

A top obedience contender and tracker, as well as a great companion, Toby and I soon became regulars in area dog shows and, in spite of my inexperience, he won a number of ribbons and honors. He was on the verge of obtaining recognition as a bench champion when kidney illness took him in 1948 at the age of seven, an enormous loss for me.

Although I could always count on Toby for reasonable performance, I always wanted to learn more about training and showing. I well remember, after his first dog show in this area, in which he won Best of Breed, I phoned the professional handler whose dogs we had edged out. I told him how much I liked his dogs and wondered if he had any time to advise me on handling, as he himself was considered the best handler in the area. His reply was brief, "You seem to be doing all right by yourself." I guess he just didn't like being beaten by a rank amateur at the end of the leash, overlooking the fact that it was Toby, not me, that had taken the ribbon.

Eventually we became friends, but in the meantime I turned to another handler whose name was Bill Trainor. Bill couldn't do enough for anyone, as long as the quality of the dog was uppermost. At dog shows, he believed, owners should not try to hide faulty traits but emphasize the good ones.

Goldens for Hunting and Conservation

I mentioned earlier that it was Mark's love of duck hunting that opened my eyes to the attractions of Toby's breed. As Mark's companion, even at the age of one year, Toby amazed other duck hunters with his willingness and ability to retrieve.

Toby became one of the first Goldens to be used in the marshes along the Maine coast for conserving waterfowl and upland game birds. Using retrievers for conservation began in the 1940s. It involved having retrievers find and recover downed birds. In the decades before the use of retrievers became popular, hunters might lose as many birds as they brought to hand simply because wounded birds flew out of reach, fell in heavy cover, or landed in deep water.

Featherquest Jack Daniels, "Danny," early 1950s.

As Mark and Toby demonstrated to hunters along the Maine and Massachusetts coasts, the loss of birds is dramatically reduced by hunting with a retriever.

Though I loved training retrievers in the fields around River Road Farm, sitting in a duck blind in a cold marsh was not my idea of a good time. I vividly recall a particular dark morning when we were crossing a channel to get to a duck blind up the marsh. The tide was rising, and the top of my rubber boots were just short of the water's depth. Since Mark was carrying the guns and a string of decoys, his rugged partner, Ray, offered to carry me across the channel. Mark made the other side safely, but Ray found me heavier than he had figured and dropped me halfway across. My boots quickly filled with the rising tide, but we went on.

Featherquest Rocky at nine months.

The highlight of that morning was when our four-legged hunting partner, Danny, a son of Goldwood Toby, took off after a bird that had been hit but not dropped and had flown out of sight. Chasing it around a bend in the channel, Danny was gone so long we began to worry. Twenty minutes later, he reappeared, tired and muddy, carrying the injured bird which we quickly dispatched. Had Danny not recovered it, the bird would have faced a lingering death, offering an impressive example of the use of retrievers in the conservation of game. Danny was as talented as his sire, Toby, and became the first New England-bred retriever to qualify for so-called "limited all-age" field trials, winning first place over 44 other dogs.

Pursuing our interest in this special use of working retrievers, Mark and I sought statistics from game and conservation organizations across the country to determine the average annual loss of crippled birds not accounted for, or shot birds never retrieved by shooters. The figures we obtained were staggering and my interest in training for the show ring expanded into training retrievers for the field.

Recognizing our concern, Ted Rehm of Taramar Kennels on the North Shore, invited experienced trainers from Long Island, New York, to come to River Road Farm for a daylong demonstration with their own working retrievers. As a result, we began holding training classes at the farm, and in 1948, Mark and I, with the help of other enthusiasts, established the Colonial Retriever Field Trial Club. Mark was the first president. Owners of Goldens, Labradors, Flat-Coated and Chesapeake retrievers participated in the training classes and we had good fun as we spread the word about this marvelous breed. Goldens predominated at first, but as time went on Labradors became more prevalent.

I recall one lady who raised Labradors and joined these training sessions. One day the class instructor privately expressed his frustration to me that her dog appeared to make little progress from one session to the next, which he feared was holding up the class. When I brought this up as delicately as possible, the Lab owner burst out laughing. "I guess he doesn't realize that I'm bringing a different dog each time!"

Colonial Retriever Field Trial Club at River Road Farm 1950. I am 2nd and Mark is 4th from the left. Seven of the twelve dogs are Featherquest.

Toby also a played a large role in my interest in the gait of dogs, an interest that blossomed into years of research, writing, filming and speaking on the topic. Once, at a dog show, I was surprised when the judge stopped me as Toby and I were leaving the ring. "Miss," he said, "I just want you to know your dog moves beautifully. A glass of water wouldn't spill from his back." Someone else had said the same thing a while earlier. Each was referring to Toby's smooth working top line or back which is essential to a dog's sound structure and contributes to endurance. But why did some dogs have it and not others? In time I found out why.

It is said that, "The eye sleeps until the spirit awakens it with a question." Over the years to come, my inner eye would become quite awake searching for the clues to this puzzle.

Featherquest Champions over the years. Art by Marsha Schlehr, 1993.

Chapter 7

WRITING DOGSTEPS

By the time Mark and I returned to New England after the war, Goldens were gaining a foothold in the East through bloodlines new to me. Imported from England, they were a bit heavier in bone, squarer headed and lighter in color than dogs from Toby's Midwestern line, which had come down earlier through Canada. I swallowed hard when a stranger once mistook my beloved dark Golden Retriever for an Irish Setter. Some time later, when someone else jokingly compared Dennis, a Golden we had bought from a California breeder, with a small St. Bernard, I reasoned that since Toby and Dennis retrieved ducks equally well, the differences were just another interesting feature of the breed.

The idea that such variations could begin to undermine the basic qualities of the breed occurred to me and others about the same time. My wake-up call came around 1950 at a large dog show in which a light-colored, 27-inch-tall Golden Retriever male took the Best of Breed ribbon, while a dark female Golden Retriever, only 20 inches tall, prevailed among dogs of the opposite gender. Dog shows customarily offer both "Best of Breed" and "Best of the Opposite Sex" ribbons. As I watched at ringside, I overheard a bystander asking a friend if the winners were the same breed of dog! I thought she was joking. She was not. Shortly after that, an advertisement appeared in a popular dog magazine describing mats large enough for 27-inch dogs—which is three to four inches taller than a typical Golden Retriever. The photograph showed an enormous, blissfully contented Golden occupying one of the mats.

Tennessee Jack Daniels "Danny" field training in 1951.

With these warning flags waving, it was time to bring the various sides together and agree on the characteristics of the breed. With the support of Reinhard Bischoff and other influential members, we stirred the Golden Retriever Club of America, or GRCA, into action. A committee was appointed in 1950 and turned over to me as chair to clarify the breed standard.

I had by that point become increasingly active in the GRCA on whose board I would later serve. In fact, in due course I held every major club office but treasurer (heaven forbid!). I was motivated by my growing interest, as far back as our time in Indiana, in promoting good-looking working Goldens. Mark shared that goal, which is why we began holding informal breed matches soon after buying River Road Farm.

Ch. Featherquest Nugget flying into the water.

My committee sought to develop breed standards for height and body proportion that would stress moderate size as well as sound structure and movement. More ambitiously, we set out to describe in detail normal versus faulty gait. The breed standard has been further refined since then as judges have shown more interest and as the need for education has become more evident.

It was this work defining the nature and purpose of Golden Retrievers that focused my interest on dog locomotion, setting me firmly on the path that would lead to the publication of my book *Dogsteps* in 1973 and my long lecturing career. Or at least that was the immediate impetus. In the back of my mind was that comment the judge made years before praising Goldwood Toby's sound, balanced structure and smooth way of moving. Why, I wondered, was Toby like that? And why were some of the many horses I'd ridden in my life more comfortable to ride than others, especially on long rides? My curiosity was about to be put to the test!

A Filmmaker is Born

So it was that I began taking movies of dogs in the early 1950s with an old 8mm camera. I did this partly to help our puppy customers understand what they should be looking for in dog structure and to help clarify normal leg movement. The movies, played back in slow motion, fascinated me and deepened my interest in the complex interplay of bone and joint structure in a moving dog. I made these movies over a period of 30 years, upgrading my equipment as time went on, eventually turning, as I explain at the end of the chapter, to moving x-rays, or cineradiography, to obtain an inside perspective on dog anatomy. Equine friends followed my work with interest, as they too were well aware of similar issues in the field of horses.

With my trusty 16mm Beaulieu movie camera that I used to film the dogs for lectures and to produce Dogsteps.

Although I was an amateur behind the camera the early results were eye-opening, and interested local groups soon asked me to present talks based on my films. One such film showed a team of Siberian Huskies taking off at a brisk trot during a race and breaking into a gallop. I've always believed that observing all types of dogs helps one to better understand one's own breed. As they approached the finish line, the heads of several Huskies could be seen drooping and their bodies moving awkwardly. Such irregular movement might be a result of lack of conditioning, but it could also be caused by faulty structure, which forces the dog to work harder and thus to tire more quickly. I shot another sequence from a bouncy rig being pulled by a trotting team. The rear limbs of the right-wheel dog seemed to buckle as they pulled the heavy load, signifying a serious problem in the dog's pelvic structure, possibly hip dysplasia.

Around this time I read McDowell Lyon's classic 1950 book, *The Dog in Action*, the first on this subject, I believe, to be published in this country. Though I later came to disagree with him on some points, I recommend it as a thoughtful addition to anyone's library. I wrote Mr. Lyon about my film study and our committee's description of gait. He answered immediately, saying we were on the right track and that he hoped to stop at our farm when he returned from a judging assignment in Canada. I looked forward to meeting him, but unfortunately he took ill and passed away before that could happen. I then turned to Lloyd Bracket, a respected dog authority who wrote a series of articles on structure and gait in the magazine *Dog World*. Unfortunately he, too, passed away before we could meet.

Undaunted, I then sought the help of Laurence Horswell, an author and specialist in Dachshunds who knew of my work. Mr. Horswell visited the farm to see the 8mm films he had heard about and became a supportive friend for years to follow. He appreciated how the zoom lens of my camera kept a dog in focus from the near to far side of a ring and back, maintaining the same approximate image size. The lens could also limit the frame to the dog in order to avoid identifying the handler. This was particularly sensitive when faulty gaits were recorded since filming a handler would likely result in identifying the breeder. Anonymity of handlers and breeders ensured better access to photo subjects, both good and bad, for my ongoing study. Playing the films on my projector at half-speed made the gaits easier to study. And the projector's stop-action helped, especially when using reverse mode to make the dogs move backward.

In 1969, the 8mm film I had been using wore out. This was all the more regrettable because I had no back-up copy, and I recognized that it was time for a 16mm movie camera and a projector of higher quality. With my husband's kind support and generosity, I bought a 16mm Beaulieu camera with a 12-to-1 Agnenieux zoom lens. This lens was state of the art at the time and allowed precision zooming, great flexibility in filming under natural light conditions, and unparalleled clarity of image. I also purchased the necessary editing equipment and a professional projector that could slow down, speed up, stop or reverse the action. Being a rank amateur, I tossed more film into the waste basket than appeared on screen, but I've learned that even professionals often throw out just as much.

From time to time, friends interested in my work brought dogs for me to film, or I would travel to film them herding and tracking in the field or at obedience events. At dog shows, I would some-times notice exhibitors shying from "the lady with the zoom lens." I did keep from identifying handlers whenever possible and often asked for signed releases, which were gladly given. It took time and effort, but over the years I compiled seemingly endless 16mm foot-age of sequences of canine locomotion of seventy or more breeds. My movies even prompted some viewers to study the leg action in human joggers along the public roads! I especially enjoyed watch-ing marathoners approach the finish with their legs outstretched in powerful strides, always reaching toward an invisible center line to maintain balance. In dogs, this is called the tendency to "single track."

From Reels to the Printed Page

When viewers attending my talks began asking for material to take home, fate gave me a boost in 1969 in the form of a letter from Ellsworth Howell, founder of the Howell Publishing Company, a major dog book publisher, asking me to write a book on the Golden Retriever. I declined, one reason being that a good friend, Gertrude Fischer, had begun work on a definitive Golden Retriever book ten years before for which I had contributed a chapter on the breed standard, illustrated with early photographs pertaining to the origin of the breed from my own collection. Gertrude's book had not taken shape because of a contractual dispute with her original publisher, and I knew that she would be happy to find a new one. Knowing she was an able writer who had many contacts, I urged Mr. Howell to pursue working with her. He did, and her book became one of his best sellers.

A more compelling reason I declined the offer was that I wished to try my own hand at a book specifically on structure and gait, illustrated with line drawings based on sequence frames in my 16mm films. Mr. Howell had attended one of my lectures, and when I sent him a sample of my sketches, he jumped at the idea and encouraged me to proceed with the project. So began four years of preparation that culminated in the publication in January 1973 of *Dogsteps. Illustrated Gait at a Glance.*

The basic plan for my book was to select multiple short series of frames from the 16mm footage showing various sequences of footfalls, each sequence illustrating variations in normal or faulty movement. For each series, I would trace the images of the dog selected to illustrate a particular feature, with brief subheadings and explanations beneath the drawings. My LW-224 projector, which allowed me to focus easily on individual frames, provided the necessary accuracy for the results I hoped to obtain.

The next question was how to simplify tracing the images. Tracing anything on an upright screen standing to one side of the projector, was not practical. I needed to have the projector facing me. Figuring out a satisfactory arrangement was a challenge, but in due time I succeeded, with the help of my artistic son-in-law. This was my solution: I had an old wooden drawing table with an adjustable slant top. I cut a 10-by-12 inch hole, over which I placed a sheet of Plexiglas. To have the projector face the underside of the table, I shortened the back legs of an old chair so that the seat slanted upwards, and I then placed the chair a short distance behind the drawing table. I then set the projector on the slanted seat with the lens pointed directly at the opening we'd cut through the top. The zoom lens and stop action brought the selected frames into sharp focus through the transparent drawing paper covering the opening—a satisfactory arrangement for tracing the hundreds of images to follow.

As the number of strips of sketches grew, showing the sequence of footfall from every angle, both correct and incorrect, I began taping the strips to the walls of my study. The room soon resembled a gaiting frenzy, but using the walls proved to be a great idea for selecting which of my many hundreds of illustrations were the most meaningful. More than 450 of these were used in the final book. Selecting sequences of individual frames from the 16mm film, tracing the images, and finally creating the pen and ink illustrations, often took countless hours over many weeks with revisions, re-tracing, and finally conveying the desired illustration.

With the text completed, it was time to decide on a dedication. It wasn't a difficult decision, for who could it be other than Mark, who had read and re-read so many pages, and listened patiently while sitting by my side for hours in the film studio? The professional film editor certainly appreciated his being there, since he needed my husband's help in persuading me that certain sequences were repetitious or unnecessary. Mark's dental practice also provided crucial financial support to my book project and when he read my tribute, "To my husband, whose patience made it possible," his response was witty and humbling. "Thank you, honey," he said, "that's very kind, but you've spelled *patience* wrong."

Dogsteps won the top award from the Dog Writer's Association in 1974, the same year I was honored by receiving the Gaines® Award for Dog Woman of the Year. With help from Ellsworth Howell's promotion, the first edition went through eight printings and *Dogsteps* became a best seller. The awards were an honor, of course, but I was gratified even more by support and praise for the book from dedicated dog breeders who joined me in the need to promote sound breeding programs based on facts, whatever the purpose of the dogs they raise. And though I at last had completion of the book behind me, the last sentence about structure and gait, as it were, was yet to be written—as I would soon learn.

An X-Ray View of the Dog in Motion

My interest in gait took a new turn after *Dogsteps* came out, when I agreed to help the Golden Retriever Club of America address the challenge of hip dysplasia. The need for action had been stressed by my friend Vern Bower, who believed that breeders needed more information about the importance of breeding away from this often crippling abnormality. She was the driving force behind the creation of the GRCA Council on Hip Dysplasia.

My own veterinarian, Dr. E.W. Tucker, a former president of the American Veterinary Medical Association, stirred the national association into action. A national meeting was called and an illustrated booklet was published on the subject. As other breed clubs became involved, the council was absorbed into the new Orthopedic Foundation for Animals, and I was named to its board. The foundation was started by John Olin of the Olin Corporation (a manufacturer of copper alloys). John himself suffered from hip dysplasia as did one of his Labrador Retriever field champions.

I had for some time been curious to see what went on in the moving pelvic assembly in dysplastic dogs and was trying to figure out how this could be photographed when an unusual chance came my way at Harvard University's Museum of Comparative Zoology. It had a specially designed fluoroscopic lab for filming animals the size of large dogs as they moved on a treadmill. The technique is called cinefluoroscopy, or cineradiography, and results in 16mm black-and-white film showing moving bones and joints, like moving x-rays.

Professor Farish Jenkins, an anthropologist and head of the museum, had designed the equipment. He became aware of my interest in canine movement and asked if I might supply him with a few dogs for his own study. The greyhounds that he had tried to use could not adapt their long, coursing gait to the treadmill. I was only too happy to do so and had no trouble finding subjects for him, thanks to friends who shared my interest in the project. In return, Dr. Jenkins gave me the opportunity to study different kinds of dogs on my own. I think he opened the door a little wider than expected, as I found dog after dog whose inside structure I wanted to see. Each trip to the museum required special arrangements to reserve the lab in between student use and involved x-ray technicians who could handle the lead-encased camera on one side of the treadmill, the fluoroscope on the other, and could at the same time control the varying speeds of the treadmill.

The slight time during which the dogs were exposed to radiation on the treadmill was considered harmless, but the technicians and I were always draped in heavy lead aprons. Whether being a Radcliffe graduate helped or not, I don't know, but the museum staff was very cooperative and I shall ever be grateful. Dr. Jenkins would often refer to my breed selections as "Pagey's canine exotica." The techniques we were using were very new, so it wasn't surprising when people were confused as to what we were doing. Once, I attempted to explain all the *x-ray* filming to a friend's daughter. When I saw her mother soon afterward, I discovered my effort had failed miserably. "My daughter wanted to know," my friend said laughing, "what a nice lady like Mrs. Elliott was doing making x-rated movies."

My research at Harvard produced graphic evidence of bone and joint motion that irrefutably challenged some erroneous notions that had been held far too long in the dog fraternity. As a result of the moving x-rays, the serious problem of canine hip dysplasia, in its varying degrees of severity, came to light, not only shown from side view but also from above. One tragic case showed a dog with no hip sockets, the femoral heads disconnected and wobbling in the pelvic muscles. Dr. Tucker, the veterinarian and friend who wrote the foreword for *Dogsteps*, couldn't believe what he saw.

In regard to the structure of the front quarters, the cineradiography has given countless viewers, breeders and dog judges persuasive photographic proof that the long speculated (but never confirmed) notion of a 45-degree angle of the dog shoulder blade is a myth. It shows that another myth, that of a 90-degree angle in the shoulder joint, where the lower blade meets the upper arm, is also a biomechanical impossibility. Those erroneous theories held sway for a long time, regrettably never halted despite lack of evidence, and dogs were often faulted for front structure that was actually correct. The earliest challenge to the misconception that I've been able to find is a small book written in 1944 by a respected Dutch breeder of German Shepherds, H.A.P.C. De Groot. A dog judge himself, De Groot differed here with a major figure in the field, Max von Stephanitz, also a distinguished breeder, judge and author.

DeGroot took clinical measurements of more than 100 soundly built dogs that moved well, yet he never came close to finding either a 45-degree angle of the blade or a 90-degree angle at the shoulder joint. The significance of the misunderstanding lay in the fact that dogs not showing the presumed or theorized angulation were criticized at dog shows for being "steep" in the shoulder or lacking angulation in the front quarters. The tragedy of it all is that, eight years later, McDowell Lyon perpetuated the myth in his 1950 treatise on dog structure and gait, *The Dog in Action*. The book contained erroneous definitions and misleading diagrams that have been copied by others to this day. Though Lyon's book holds material of worthwhile interest, I feel sure that the author's position on angulation would have shifted had he lived to see the moving x-rays.

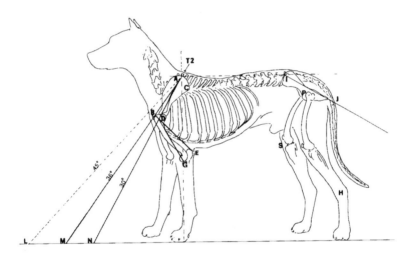

The fallacy of the 45-degree layback exposed.

Dr. Farish Jenkins' answer to all of this was clear and direct. "Of course there's no 45-degree layback of the shoulder blade," he said, when I telephoned recently to confirm his position of so many years ago. "Such an angle would be functionally impossible. The shoulder is attached to the chest wall only by muscle. Were the scapula set at a 45 degree angle, the shoulder joint would set in advance of the sternum, with no support of the fore chest, and the upward excursion of the shoulder blade would interfere with the musculature of the cervical vertebrae." The shoulder blade normally sets at a 30-degree angle off the vertical, Dr. Jenkins said, adding that many dogs move well with less. It's important to remember, he added, that the length of a dog's stride is controlled not only by the excursion of the shoulder but also by the swing of the upper arm. "The angle of the lower blade with the upper arm is normally between 108 and 115 degrees, not the 90 degrees so wrongly shown in many diagrams."

The first edition of *Dogsteps* was already out, too late to include this exciting research. But at my publisher's request, we produced a new edition in 1983 updated to reflect my work regarding the correct angle of the shoulder blade and shoulder joint. The second edition, *The New Dogsteps. A Better Understanding of Dog Gait Through Cineradiography*, included many new sketches based on my research in bone and joint motion through the use of moving x-rays. In 2001 a third edition was issued in paperback and is still available as *Dogsteps. A New Look*, which further emphasizes the importance of sound structure in the strenuous sport of agility.

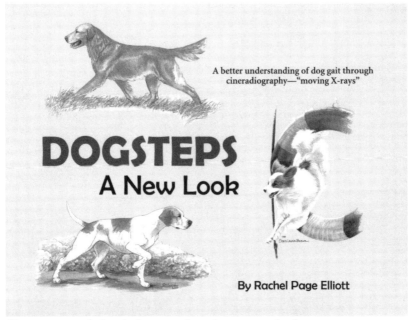

Dogsteps opened the window on gait and movement in dogs.

Encouraging letters of appreciation and news clippings gave me strength, in addition to my husband's support, to continue giving my illustrated film presentations. I recall in particular a letter from a Swedish writer, Kerstine Bermings, who enclosed a translation of an article she wrote about how much Collie breeders appreciated learning the truth regarding shoulder structure. She wrote how the breeders feel that *Dogsteps* is helping to replace misguided criticism of what has always been normal structure with acceptance and understanding. "We must remember that the history of science is filled with examples on how old theories have blocked the road for new discoveries and findings," wrote Ms Bermings. "As soon as old theories are refuted, one is able to see new things in dog movement. Look at your dog with fresh new eyes, and you'll see things you have never seen before simply because misguided theory has forbidden you to see it. When the myth about the new clothes of the Emperor was cast down, everyone could see that he was naked!"

Mark and me with Stormy in 1965.

Chapter 8

HAVE PROJECTOR, WILL TRAVEL

With the publication of *Dogsteps*, I became an author at age 60. That was satisfying, but it wasn't the last "new trick" this old dog-lover would learn. As invitations to speak about canine structure and movement grew, I stepped into a new role, that of professional lecturer. Using films and photographs to illustrate my points, I lectured for over two decades before breed clubs, dog societies, research groups and veterinary associations on three continents (North America, Europe and Australia). Enthusiastic audiences included breeders, judges, veterinarians, professional trainers and handlers plus, of course, interested owners.

I actually gave my first lecture in 1969, four years before *Dogsteps* came out. On the recommendation of my friend and supporter, Mr. Horswell, the Dachshund specialist, I was asked to address the Professional Dog Handlers Association's meeting in Maryland. The exposure from that talk and the release of *Dogsteps* increased the number of requests for me to speak. But it was the cineradiography of dogs on treadmills that really pushed my lecturing into high gear. The fluoroscope-based images showing the bone and joints moving "inside" a dog in motion added a striking new dimension to my research and standard movies.

I made my first international tour in 1976, under the auspices of the Golden Retriever Club of England, the first of three lecture tours in England. My hosts and I had good times and much laughter, as well as long drives and weary nights following meetings. Not only was the tour professionally satisfying, but I was enriched by visiting famous old kennels and making many new friends. I well remember one event where we found the audience queued up outside the meeting room because no one had remembered to bring the key. The key was sent for and retrieved, but with a noticeable lack of haste. No problem. The British never seem to be in a hurry.

The technical success of my talks depended on my LW-224 projector, which I bought in 1969. It had various speeds, smooth projection and instant forward or reverse. I could operate it by remote control, which was handy during a presentation, and freeze the action without scorching the film. I carried it with me in the United States, as I did not expect to find one in local areas. When I traveled abroad, however, I dared not check it through as passenger luggage. On two occasions when I relented, it was damaged.

Not even my trusty projector was immune to problems, as I learned at a meeting of British veterinarians. As I was showing a movie, one of my projector's take-up reels broke down and began spilling film onto the floor. Despite the darkness, someone noticed the film mounting in great loops behind me. It took some effort, but I managed to rewind the tangled film and was forgiven for the delay in the convention schedule. The reward for my troubles was seeing the veterinarians' fascination with the moving x-rays and sensing their appreciation of the message I conveyed, particularly for students in their profession.

I don't think anyone described that message more clearly than my friend, C. Bede Maxwell, an international dog judge and writer. Once, when she was visiting me at River Road Farm, I took her to the Museum of Comparative Zoology at Harvard to watch the making of a moving x-ray. Her excitement knew no bounds, and she lost no time sharing it through the British weekly *Dog World*. Her article, headlined *An Amazing Technical Achievement*, appeared August 11, 1978.

"This newest miracle in an age become blasé about miracles should excite every dog owner, breeder, judge," she wrote. Combining x-ray imagery and filmmaking, she noted, has enabled a visual depiction "of the actual movement of parts within the body of a subject animal as it moves steadily on a treadmill it has been schooled to negotiate." Having watched the results on a monitor, she noted: "The initial experience of seeing [the result], which comes to the screen as a 'positive' x-ray film in tones of soft sepia, is mind-blowing. There, clearly outlined in movement, is the shoulder blade, the upper arm, the elbow joint and, in back, behind the lattice of the ribbing, the pulsing of the heart, making clear the engineering process that has always been so difficult for newcomers to the world of dogs to understand."

A Civilized Quarrel in England

A second trip to Great Britain in 1978—I was beginning to feel at home there—fell under the auspices of the World Congress of Kennel Clubs meeting in Edinburgh. I upset the applecart by insisting on a screen with better reflective quality than the one provided. In order to show the cineradiography to its best advantage. This caused a delay and being a Sunday, all the rental agencies were closed, and the chairman was quite unhappy. But they somehow managed to find the radiant screen I needed and the presentation proceeded.

That trip coincided with an invitation to judge an open Golden Retriever show in Wales. It had 225 entries, making it the largest show I had judged. But with help from many friends I managed to cope with a classification system quite new to me. One lovely dog, Nortonwood Checkmate, caught my eye from the start and was my top choice. Not yet a show champion, he won that title shortly afterward. I was sorry I could not take him home with me. However, two years later, when visiting Madge and Ron Bradbury, owners of the Nortonwood Kennel in England, I did find a daughter of Checkmate, named Nortonwood Carna. She became a delightful addition to the family and an excellent producer.

A stimulating acquaintance made during my visits to England was Tom Horner, another judge and the author of *Take Them Around, Please* (now out of print). After attending one of my seminars, Tom challenged my assertion that dogs have a natural tendency to single track. This tendency, which is also called "limb convergence," refers to a dog's natural predilection to reach slightly inward but with each forward stride toward a center line of support.

Nothing could dissuade Tom from the opinion that the legs should move parallel with the feet landing as far apart as the elbows. Not even cinematic evidence of footfall sequences played back in slow motion would convince him. He even criticized the dogs I used to illustrate normal leg convergence, saying they had faulty structure! Admittedly, the degree of convergence varies with speed as well as breed type, but all dogs make the effort in order to maintain balance and counter any side-to-side shifting as their speed increases.

It seems that Tom, in spite of the evidence, failed to recognize the true position of the shoulder blade, which is attached by muscle only on a tapering, inward slant to the elliptically-shaped forward part of the shaped ribcage. The swing of the shoulder blade initiates the inward reach of the forelimbs, along with the humerus or upper arm, as nature intended. Curtis Brown, a friend who has also written on the subject of canine locomotion, told me that he once showed Tom still photographs of trotting dogs that illustrated correct leg convergence. Curtis said he was amazed by Tom's unequivocal reply: "Even cameras can be faulty."

Tom wrote several articles challenging my presentation, albeit with somewhat courteous restraint that was absent in his attacks on McDowell Lyon's book *The Dog in Action*. I responded by explaining where I felt he had misinterpreted parts of my talk, doing so in the spirit of friendly confrontation. One of my friends wrote a limerick capturing the spirit of our lively disagreement, which I cannot resist sharing. It is entitled:

The Ballad of Tom Horner

Mr. Tom Horner stood in his corner
Watching the little dogs gait.
He fumed and he fretted,
And plainly regretted
That none of them trotted quite straight.

"Take them around, please," said Tom with a frown.
Around," he continued to urge,
So they trotted quite wide, stride after stride,
For no dog would dare to converge!

Now outside the ring, sitting neat as a pin,
Was Rachel Page Elliott, the "pro."
She squirmed, she squiggled, to herself she did giggle
And thought, "My, what he doesn't know!"

So after the class, up stands this brave lass
And spoke, before good friends could warn her.
"You surely must know that dogs' legs won't go
Like pendulums, dear Mr. Horner."

But as Tom was the judge, he just wouldn't budge,
His confidence quite unaffected:
"My dear, you're quite wrong for, in moving along
The legs are parallel—just when collected."

But lest you should fear, let me make myself clear
And do not have doubt in your mind;
Opinions on gaiting are most stimulating,
—No two better friends could you find.

From Lecture Hall to Hospital

In 1978, I went to Oslo, Norway on invitation from the local Golden Retriever Club. For all the x-rays of hips I had seen over the years, I saw one in Norway that I never expected to see—my own! Blinded by some stage lights I tripped and fell from a stage when leaving the platform after my opening remarks. Funny thing about broken hips, they hurt if you move, but by sitting absolutely still beside the projector and moving only my hands to control the varying speeds and stop action, I was able to complete the two-hour program. Members of the audience who had seen me fall were far more worried than I. It was midnight by the time four red-coated paramedics carried me down three flights of stairs (the elevator wasn't working) and off to the hospital. I had managed to break my hip but I did not want to disappoint the audience.

The language barrier was a problem at the hospital, but I somehow persuaded the lab technician to let me see the x-ray. When I exclaimed, "At least I don't have hip dysplasia!" formalities were dropped and those on the hospital staff who shared my fondness of dogs became attentive friends during my 14-day stay. The prolonged visit in Oslo was not too easy for my hosts, Diane and Ray Anderson. I will never forget their gracious concern and care, nor how comforting it was, with my husband so far away, to find their personal physician by my side as I came out of the anesthesia following the hip operation.

The sympathy card I received from Kathy Liebler, who had been with me in England, was brief and to the point, "I knew you couldn't get along without me!"

Lecturing Down Under

Two years later I traveled to Australia to make 11 presentations across the country under the sponsorship and planning of the Royal Agricultural Society of New South Wales. I had heard of Australia's reputation for unmatched hospitality but was unprepared for the marvelous experiences that awaited. Judges from all over the world participated in the magnificent spring dog show held at the Royal Agricultural Society's showground. I will also long remember my tours of the Taranga Zoological Gardens and the Sydney Opera House. These tours were conducted by my affable host in Sydney, Jann Trout, who was the breeder of some of the most consistently beautiful Goldens I have seen.

Other highlights of the trip were watching the Spanish Andalusian Dancing Stallions perform and visiting a farm where Australian dingoes are raised and trained for obedience. The family that runs the farm hopes to have these unique animals accepted in today's world of dogs and ultimately recognized as a breed.

After my first talk in Sydney, one of the judges, who was a Japanese massage therapist, persuaded me that I should have a treatment before the strenuous travel schedule ahead of me. I succumbed to the invitation and stretched out flat on the floor with my head on a pillow. "Muscles no good, muscles no good," he muttered to himself. But I sure felt better!

The person who perhaps more deserved the massage was my tour guide Bob Curtis, whose exhausting job it was to look after me during our extensive travels across his country. We flew from Sydney into the back country, to Canberra, the capital of Australia, and to Tasmania, Adelaide, Brisbane and points in between. The visit to Adelaide, the capital of the state of South Australia, was memorable as busloads of judges, exhibitors and breeders from across Australia

Meeting a beagle in Alburg, Australia, early 1980s.

swelled the ranks of local dog people. Hundreds of people packed the large hall where I spoke. I couldn't believe it when they rose as one to give me a standing ovation at the end of my presentation.

Equally memorable, albeit for a different reason, was a long, bumpy and much-delayed flight of three hops from Tamworth to Coolangatta, on the eastern coast of Australia where the states of

Queensland and New South Wales meet. Our small Cessna jet took off into a raging rain storm, which shook it and did not abate. At our first stop, the airplane's brakes jammed, and before departing again the male passengers were called upon to rock the plane back and forth in a pouring rain until the brakes became unstuck. Then, airborne again, the weather suddenly closed in so heavily that the pilot asked for our help in spotting either mountains dead ahead or a break in the clouds through which he could pass and land the plane. When he finally succeeded in doing just that, Mother Earth never felt so solid or so good.

Despite the white-knuckle excitement, Bob's thoughtful attention in general took the worry out of traveling and afforded me a needed chance to rest between lectures. A special treat that Bob arranged was our visit to a large Connemara pony farm in the sharp, rugged hills of Queensland. The Connemara is a rugged, large pony or small horse, originally from Ireland, useful for cross-country riding, jumping, driving and even farm work.

A Golden Pilgrimage

I made my third visit to Britain in 1982, taking part in a series of seminars from southern England north to Scotland sponsored by Pedigree Chum®. Our guides, Eric Smethurst and Bill King, planned the itinerary in detail. Kathy Leibler, who was again my travel companion, and I were even given a mobile intercom hook-up in case we ran into trouble as we drove from place to place.

The crisis on that tour was a major one. It happened in Glasgow where, according to schedule, we met our guides Eric and Bill for the last seminar of the tour, only to find them in deep despair. All of the equipment for the seminar, featuring me and several other speakers, had been stolen from the transport car. Two projectors, booklets and advertising materials were gone! Their greatest worry, however, was about my films. Fortunately, it was my practice never to let anything so essential out of my hands when traveling and so they were safe with me. There was no trace of the stolen goods, but with their usual efficiency the Pedigree Chum staff located replacements, including the only other projector in England with the special features I needed. It was flown up from London in time for the seminar the next day.

I must add that Bill King's attention to my personal needs was heartwarming. In the middle of my lecture was a long coffee break, during which I decided to leave the buzzing crowd and find a room in which to sit and rest. I told no one where I was going. My rest turned into an unexpected nap. When I awoke, there was Bill sitting in a nearby chair, who explained that at no time did he dare let me out of his sight.

Seeking the Origin of Goldens

Kathy and I made the most of every stop to visit Golden friends wherever we could. One was Elma Stonex, whom I had come to know and cherish as a friend during a ten-year correspondence that began as a result of our mutual interest in the history and origin of Goldens. Like many other people, I had always accepted and loved the story that the forebears of the Golden Retriever breed were Russian circus dogs. But in the mid-1950s, my friend Leila Sears showed me an article from a British magazine disputing this story. The writer was a descendant of the English aristocrat and dog lover, Lord Tweedmouth, who wished to set the record straight. He wrote that he had found among Tweedmouth's papers a pedigree, dated 1868, of the first recorded litter of so-called Yellow Retrievers. It listed a now extinct breed, the Tweed Water Spaniel, as the mother of the litter and a Yellow Retriever from a litter of black puppies as sire. I sent a copy of this article to Elma Stonex. After reading it and confirming its accuracy with the author, Elma, much to her credit, halted the printing of her new breed book, *The Golden Retriever*. She knew it was important to include this new and revealing information.

Knowing little about the Tweed Water Spaniel, Lord Ilchester, the author of the 1952 article, asked anyone who came across information about this dog to contact him. I resisted dispelling the Russian legend, but I felt challenged to search for more information. Having inherited my father's obsession with old books, I right away searched through the dark and dusty recesses of a nearby public library. There I found an early 19th century book on dogs by Hugh Dalziel, in which the Tweed Water Spaniel was described as one of three types of spaniels. I sent this reference to Elma Stonex, although I assumed she was already familiar with it.

She was not, and her appreciative response was the start of a long and informative correspondence. The letters are now a part of the Golden Retriever Club of America's archives, along with an original oil painting of a dog identified as a Tweed Water Spaniel, which I bought from Gerald Massey, the renowned world authority on sporting prints and books.

Elma became a close and valued friend, sharing early photographs, prints and other records of the Tweed Water Spaniel. We also shared our interest in horses. I treasured the visit Kathy and I were able to make on that tour to her home in Taunton, England, one reciprocated when Elma later visited us at our farm. An experienced horsewoman herself, she enjoyed hitching our Morgan horse, Tempest, to a cart and taking the reins for a drive, often along the road in front of our house. Only one problem—my British friend needed several reminders that here in the United States, we don't drive on the left side of the road! Tempest did his best to remind her as well and was probably more effective than I in that regard.

Tales of Guisachan

My 1982 visit to Taunton also served as an inspiration for a kind of pilgrimage Kathy and I undertook after leaving Elma. We decided to head north toward the Shire of Inverness in the hope of locating

Guisachan Estate, the ancestral home of the Tweedmouth family, where lay as well the roots of Golden Retrievers. Our British friends did not seem to know the exact location, although Elma had said it was near Beauly. By luck Kathy came across a paragraph in an old travel book for hikers that pinpointed the location and it wasn't long before we were walking on those hallowed grounds.

One of the farmers greeted us outside the ancient sprawling stone barn and directed us up the lane to the crumbling remains of the Victorian mansion where Lord Tweedmouth and his family had entertained sportsmen, political dignitaries, artists and other members of the aristocracy. Royalty was no stranger to them.

Of particular interest was an octagonal-shaped building called the milk house, designed after a similar structure at Windsor Castle. In the center of the structure was a fountain, long since dry, that once served as a cooling agent for milk and cream. At the time of our visit, the milk house was being used for the storage of broken furniture and miscellaneous household goods. The high upper wainscoting was lined with dust-covered paintings and old prints. One print in particular caught my eye, even though I could only see part of it. That visible part showed, next to a black horse, a retriever-like dog that could have been a progenitor of today's Golden Retriever. Fascinated, I held a rickety stool for Kathy to stand on while she snapped a picture of this print.

After doing some research on my return home, I wrote the caretaker of the estate, Donald Fraser, expressing my overall interest in Guisachan and especially the print, because of its possible connection with the historical origins of the Golden Retriever. I identified myself as the current president of the Golden Retriever Club of America and explained how much interest there would be among our members. Would he be interested in parting with this treasure? Mr. Fraser kindly answered with a letter describing the print as a lithograph made from an 1839 painting by Edwin Landseer, the central subject being Victoria and Albert at Windsor castle. He said the painter visited Guisachan as the art teacher for Tweedmouth's daughter. To my great surprise and pleasure, Mr. Fraser then sent the print to me as a souvenir of my visit. To this day it hangs on the wall in our living room.

A major discovery in Golden Retriever history was this 1839 Edwardian Landseer print. Note the golden on the right side.

Although I continued to lecture in the United States for a few more years, I gave my last overseas lecture in 1984 at the International Congress of Kennel Clubs in Amsterdam. There I had the pleasure of sharing the podium with Quentin LaHam, a specialist on dog anatomy and breeding dogs and the foremost German Shepherd Dog judge in North America at that time. We had become friends over the years at various seminars in the United States and our thoughts regarding structure and gait were in agreement. Because of the emphasis we placed in Amsterdam on the mechanical impossibility of a 45-degree shoulder layback, a writer for a Dutch dog magazine who heard us speak withdrew an article he had just contributed to a national magazine in order to update it with this new information. After the congress, friends sent me English translations of DeGroot's slender 1944 book offering confirmation of our view.

Daughter Betty accompanied me on that last trip abroad. She had agreed to look after her doggy mother provided we could take a detour on the way home through Ireland to visit Connemara pony farms. We did just that, and what a delightful detour it was, making new friends with another common interest, and reestablishing our faith in this wonderful breed, not too well known in America. With the help of her daughters Cindy and Rachel, Betty and I continue to raise Connemara ponies here at River Road Farm.

Time to Slow Down

The time finally arrived when it was advisable to slow down in my traveling. The airlines were becoming crowded and I was no longer permitted to take my valuable 16mm projector on board. This model was not available in rental agencies so it had to go with me as cargo. On three occasions it was badly damaged, interfering seriously with my presentations. Needless to say, Mark was getting tired of driving me to the airport! So, after more than twenty years on the lecture circuit, I decided to transfer the lecture material to video.

Dogsteps becomes a best selling DVD.

For two years I worked with a patient professional editor who in the process learned more than he wanted to know about dogs! With the financial backing of the American Kennel Club, the program was finally restructured and updated, still under the title *Dogsteps*.

I had already produced three other videos, one for the Collie Club of America, one for the Cairn Terrier Club and another for the GRCA. Having learned what was involved in the making of a video, both in time and money, I hesitated to take on the project for the Golden Retriever Club, especially because the idea had met with some resistance from a few members. Then came a near-midnight call from Anne Shannon, whose persuasive charm caught me at a sleepy moment and my resistance was shattered. The Golden Retriever video eventually came to fruition with the help of many friends who offered their dogs to illustrate particular features, good as well as faulty. Its reception has been heartwarming and I share the credit with all who contributed to and assisted with its completion.

I later consolidated much of the cineradiography into video form for the Orthopedic Foundation for Animals. Dr. Tucker worked with me on the script and provided professional expertise in his narration of the video. That information lives on now as a DVD *Canine Cineradiography. A Study of Bone and Joint Motion as Seen Through Moving X-rays*. When this project was finally done, I received a suggestion that I do a video on the anatomy and gait of horses. My reply was that I would save that project for my next life.

Canine Cineradiography DVD came out in 2002, but was released on video in 1983 concurrent with The New Dogsteps.

Chapter 9

FITTING THE PIECES TOGETHER

Through all of these interests and projects, from making movies of dogs to writing my book and from the x-ray lab to auditoriums around the world, Mark was with me every step of the way. When I stopped traveling and returned to a quieter life at home, he sat by my side hour after hour in the film studio editing video footage. Behind the scenes has been our faithful friend and companion, Alberta White, now 82, who has lived with me at the farm for over half a century. Cheerful and ever helpful, she is loved by everyone. Years ago, we thought that bringing up the children would be her claim to fame. Then it was the grandchildren. Now Alberta has taken on a more challenging job—bringing up the grandchildren's grandmother! Without her help, my husband's loving support and the patient understanding of our children, Betty, Ruth and David, I would not have had the courage and freedom to meet the challenges and have the adventures of my life.

After so many years of preparing lectures and scurrying to the airport, it was pleasant and agreeable to spend more time at home, especially since Mark retired from his dental practice about the same time. We kept busy and were content with horses, dogs and cats, with reading more, gardening and all the attendant activities around this old farm—not to mention stepping into our role as grandparents and now great-grandparents. We also nurtured our own relationship and were grateful for the friends who influenced our lives and rewarded our years with treasured memories.

Mark stayed in touch with his short-wave radio contacts overseas, continued his Rotary work, made occasional fishing or hunting trips to Maine and indulged his love of travel with excursions overseas. I grew interested in training dogs for scent discrimination and went to field workshops in which we followed tracks laid across open fields, over walls, across roads or through woods. However, I didn't like driving far for training classes, especially in the evening. Around that time, I learned that a local dog group, ARFF (Agility is Really Fun for Fido), needed new headquarters. I told them there was plenty of room at the farm and invited them to set up their equipment in the pasture beside the house. To my delight and pleasure, they accepted my offer and I was soon involved in the sport of agility training myself.

With my mainstay Goldens, Casper and Tammy, 1999.

Sadly, I had recently lost Willow, the young Welsh Corgi I had begun to train in agility, due to an overdose of immunizations. Undaunted, I started over with my new Golden Retriever, Tammy, a daughter of Casper. Tammy became an enthusiastic pupil, despite my not being as limber as I once was. It was all I could do to keep up with her as she headed for the obstacles, trotted across the dog walk, dashed through the tunnel or cleared various jumps. It seemed that *I* was becoming slower as *she* was becoming faster. But this thoughtful and faithful dog always cooperated and waited for me to catch up, her tail wagging, as if to say, "Hurry up, old girl!" We competed until I was 91 and we still train together.

I took up agility in my 80s. I competed with Tammy at the GRCA Specialty in Rhode Island in 1999. Photograph by Hal Ungerleider.

Hosting the agility group at River Road Farm over the last twelve years has allowed me to observe the interplay between the dogs and the Connemara ponies we keep in adjacent pastures. Leaning over the pasture fence, the ponies always seemed fascinated by the leaping, climbing, dashing dogs. It was especially fun to watch the attachment that developed between a dog named "Dreamer" and a couple of the ponies. The sight of ponies and dogs in nuzzling communication is a perfect scene blending two lifelong passions of mine.

Dreamer communicating with one of my Connemara ponies.

With other members of a parish visitation committee at my church, First Parish in Concord, I sometimes visited people in rest homes or hospitals. Once I took my Welsh Corgi to see a gravely ill friend, Sally, an ardent dog lover. It was so touching to see her reach across the covers and stroke the dog's soft coat, murmuring sounds (as the nurse later said) not heard from her in weeks. As we passed along the corridors, patients who were confined to wheelchairs would offer stories of their own former pets. On leaving a certain nursing home one time, my friend and I, each of us holding a young puppy, happened to look in on a group of elderly residents listening to a solo violinist. Spotting us, the wily violinist nodded a welcome and switched to "How Much Is That Doggie in the Window?" which of course brought smiles and good cheer all around. Two friends and I regularly took our Golden Retrievers to the Veterans Hospital in Bedford, where we were always a welcome distraction with a particular group of veterans. But, truth be told, not everyone enjoyed our doggie visits, as we were once reminded. As we stepped from the elevator, we heard a loud voice at the end of the hallway wailing, "Here come the damned dogs!" Undeterred, we made our rounds, during which hands were outstretched to welcome back the dogs they had come to know through our previous visits there.

A Puzzling Passion

When one door closes, another opens. Such was the case for me when I bid farewell to the lecture podium. The door that opened returned me to one of my long lost hobbies, jigsaw puzzles. John Teele, a friend I had known through our interest in dogs and who also shared my interest in puzzles, provided an assist. John saw a few of the wooden puzzles that Alberta and I had been ordering from a company in England. Obviously unimpressed with their quality, John now and then entrusted us with choice selections from his wife's early collection.

These lovely puzzles helped reacquaint me with my childhood love of puzzles, one shared and nurtured by my father. He agreed with President Calvin Coolidge that one should "always have a jigsaw puzzle on the table inviting pause in the day's occupation," and he amply supplied our home with them. I vividly recall Father's patient encouragement, and forbearance, when we hovered over the puzzle table searching for just the right shapes to fit where they belonged. Of all the puzzles in our old family collection, my favorite showed a young colt caught in a snow storm, pawing the fence rail as he begged to be let into the barn.

It wasn't long, however, before Mark became concerned about the extravagance of my ordering puzzles from England and suggested that I try cutting them myself. Little did he know what he was getting us into! The pleasure and challenge of designing and cutting wooden jigsaw puzzles became such an absorbing avocation that I eventually started a small but rather busy cottage industry, Pagemark Puzzles.

My sister, Deborah, had already tried her hand at the craft and she gave me encouragement by loaning me a small unstable scroll saw for practice. After seeing me struggle with this unwieldy equipment, Mark surprised me on my next birthday with a Dremel jigsaw geared with a 15-inch throat from blade to post. In the years ahead more versatile machines followed, the last one was a 22-inch Hegner with greater stability, a quieter motor and a larger plate for support of bigger prints.

While the jigsaw blade is critical for precision cutting, choosing the right wood on which to laminate the picture or print chosen for the puzzle is also important. Together Deborah and I explored different kinds of plywood, including poplar, birch, cherry and mahogany. My preference became five-ply mahogany, 3/16 to 1/4 inch in thickness, which tends to minimize warping or splintering on the underside of the board and helps prevent "voids," or hollows within the layers. We never thought about using cardboard, as wood makes the puzzle longer lasting and its tactile feel is more pleasant.

I still enjoy cutting puzzles in my spare time, July 2008.

By trial and error we learned about variations in blades, too, blades measuring in thousands of an inch. Thickness of the blade controls the cutting space between the pieces, called the kerf. This varies with preference of the cutter and artistry of the overall cut. Styles differ widely, almost like personal handwriting. Some cuts are rounded, some curled, or straight edged, others angular, long and wavy, jagged or strip cut. Interlocking pieces with sockets and knobs help keep a puzzle from pulling apart. Varying styles often identify the cutter.

My own puzzles differ widely in subject matter, size and number of pieces. Cut into each puzzle are a number of distinct silhouettes, called figurals, which depend on the subject or theme of the puzzle and vary endlessly. My signature piece, not surprisingly, is a Golden Retriever.

After a puzzle is cut the pieces are inspected. Mark enjoyed helping with the finishing touches, sitting here on the living room sofa and sanding the back of each piece with an old dental tool. He said it kept his hand in practice in case he decided to go back into dentistry. One time a lost piece caused a frantic search, worrying us for two days. He found it in one of his shoes. Another time, the lost piece appeared in the cuff of his pants. The puzzle pieces are then assembled, counted and packed in a sturdy box identified with our company name, as well as the title, size, and number of pieces.

Selling or loaning jigsaw puzzles was a lot of fun, although there were trying moments. An acquaintance to whom I loaned a puzzle that I had cut accidentally left it on the roof of her car, where it was soaked by heavy rain that night. The next morning, unaware of her mistake, she drove off, sending the sodden pieces flying everywhere. It was a total loss. Another time, a regular customer of Pagemark reported that eight of the figural pieces (outlining tiny figures tied to the theme) were missing from the puzzle she had just purchased. I was devastated, not to mention puzzled myself, because I couldn't imagine overlooking such a thing. I agreed to take the puzzle back, of course, to rectify the matter. Months later, the customer, both embarrassed and deeply apologetic, called to say she found the missing pieces—exactly where she had hidden them from the children, something she had completely forgotten she had done! All was forgiven.

For a number of years I took part in craft fairs, opening yet another circle of friends. I was usually the only exhibitor of hand cut jigsaw puzzles, at that time, and often sold out by the day's end, with follow-up orders for one or two private customers. The last fair I entered was memorable. Out of forty-four puzzles I had on display, thirty-six were sold, making a good profit for the fair sponsor, Emerson Hospital, and easily covering my own expenses.

The Lure of Puzzles

Not only did the creativity of making puzzles intrigue me, but so did the history and enduring appeal of this fascinating hobby. New England has been the bedrock of jigsaw puzzle activity since the late 1700s, when settlers moving here brought various kinds of entertainment from Europe with them. To share knowledge and cutting techniques, a friend and fellow puzzle cutter, Bob Armstrong, and I invited friends who shared our interest to meet occasionally. Our "Puzzle Parley" group, as we called our soirees, eventually became part of the Game and Puzzle Collectors Association, through which I formed a close friendship with Anne Williams.

A professor of economics at Bates College in Lewiston, Maine, Anne is an acknowledged guru of the puzzle world and the author of several books on the nature and history of jigsaw puzzles. The earliest works in her own collection are from the 1760s, shortly after a London mapmaker named John Spilsbury began making the first jigsaw puzzles to occupy children on Sundays. They were not permitted to play active games on the Sabbath. Spilsbury, who owned a print shop in Convent Garden, called his creation "Dissected Maps for Teaching Geography." Each puzzle in Anne's collection is identified by maker and date. Many are on display in the puzzle museum she established at Bates.

The puzzles of Spilsbury and contemporaries generally depicted geographical maps, Biblical stories or chronological prints of royal families. Puzzles later took on an enormous range of subjects, including social disorders, war, the industrial revolution, economic depression, politics, sports and of course popular culture. And while they were originally created to instruct and instill moral values in children, they've been used (or perhaps misused) over the past 300 years to amuse, titillate and persuade.

The continuing appeal of jigsaw puzzles today is impressive. Despite our technological amusements, people continue to find the challenge of a puzzle awaiting assembly irresistible. I say this despite the dubious description of puzzle fans in the 1936 Parker Brothers game *Puzzle Pastime*. Jigsaw puzzles, the instructions noted, appeal to "weary readers," "invalids," "tired business people," "stay-at-homes" and "jaded bridge players!" Occasionally, guests to the farm have refused to go bed until the puzzle they had begun was completed. Why? One reason may be that puzzles are a metaphor for life. Like the pieces of a puzzle, people, events and activities come and go in different shapes and sizes. Some are big and some are small, yet each one in its own way contributes to the complete picture that reveals itself as part of our own life. We also know, at least intuitively, that engagement in the assembly of a puzzle keeps the mind sharp.

One for the Record Book

Upon request, from time to time I donated my hand-cut puzzles to fundraising events, including auctions for the Yankee Golden Retriever Rescue Association, the Golden Retriever Club of America and WGBH-Channel 2, the public television station in Boston. My son-in-law, the accomplished painter Maris Platais, and I occasionally coordinated our efforts, with his painting a scene—marine or landscape, in acrylic paint—directly on a plywood board, which I would then cut into a jigsaw puzzle. The combined effort of two artists always brought bids well over a thousand dollars for these original pieces.

The most unforgettable auction occurred in 2005 at a national meeting of the Golden Retriever Club of America. The ten-day gathering, held in Gettysburg, Pennsylvania, brought more than 3,000 dogs from across the continent to participate in various forms of competition—field, obedience, conformation, agility and rally. Because it was being held 400 miles from my home, I had no plans to attend. "Just tell them I'm still alive," I responded, "I'll send them a puzzle instead." However, family members and friends were unusually persistent, and in time they prevailed. I agreed to go, puzzle and all.

The gala evening and auction, which was sponsored by the Golden Retriever Foundation, was held on the last day. The men were resplendent in tuxedoes and the ladies elegant in gowns, heels and dripping earrings. I had no warning that it was black tie, but it didn't matter, as I don't wear earrings or high heels anyway. The evening began with a recap of the past year's highlights, a set of brief talks recognizing the hard work of those who had advanced the work of the Foundation, followed by the introduction of its new president, David Kinghorn.

Then came the surprise that explained the persistent encouragement I had received to attend: the announcement that the Golden Retriever Foundation had established the Rachel Page Elliott Educational Fund. This charitable effort supports scholarships at selected veterinary schools for students interested in canine anatomy and physiology, animal behavior and veterinary medicine, and it supports broader public education efforts by providing stipends to teachers, veterinary students and practitioners. Words failed as I was presented the document bearing my name.

What happened next was fun and even a bit breath-taking as the auction was held primarily to raise money for the Foundation. Jennifer Kesner, the well-known dog trainer turned auctioneer, was at her best, her humor and relentless energy evoking responses from even the most timid of bidders. Thanks to Steve Bolton, who strolled through the banquet room holding each item within sight of the 700 guests, the interest and bidding became lively. The auction drew to a close with the Golden Retriever jigsaw puzzle I had donated, starting with a bid of $500. It quickly moved to $1,500, then $3,000.

"How about $5,000?" called Jennifer. A hand waved, Jennifer nodded, and quick as a flash, it was up to $7,500 on other side of the room. "Thank you," she said to the high bidder, "and now do I hear $10,000? Over there. Yes! Now how about $12,000?"

A pause gripped the room. Then Erik Foster, who was sitting at my table, gallantly shouted "$15,000!"

Dead silence. Erik looked scared to death that he was stuck with his offer. But Jennifer released him almost immediately as she pointed to a raised hand at the other end of the room. The bidding continued, moving even higher, and the excitement reached a fever pitch.

"$17,000! Thank you. Why not twenty? Twenty, thank you! Twenty-two sounds even better? Twenty-two anyone? Twenty-two over there! *Thank you!*"

Suddenly there was a flurry of activity on the other side of the room, where a bidder was engaged in excited conversation on a cell phone. He then bid $24,000!

"Anyone else?" the auctioneer asked. "Oh, just a minute, over there! Did I hear $25,000?" Nod. "Twenty-five it is. Did I hear $27,000? Yes? Thank you!"

Again dead silence. The hammer went down and cheers shook the room. Even Mary Meador, an outstanding supporter of the Foundation and heroine of the evening, was startled as her final bid climaxed the evening.

Thrilling as that auction was for me that wasn't the end of the story. Joy Viola, the untiring development director of the Foundation,

I made and donated this puzzle to an auction in 2005 at the GRCA National Specialty in PA. It sold for $27,000 and made it into the Guinness Book of World Records.

felt we had a record breaker in an auction price of $27,000 for a hand-cut wooden jigsaw puzzle. She lost no time contacting the Guinness World Records Association in London. The response was encouraging, with the stipulation that every detail about the event had to be provided and authenticated. Guinness has stringent standards. It demanded, among other things, to know the number present at the auction and the gradation of the bids and to receive notarized letters confirming the price from both the auctioneer and the winner of the puzzle. Guinness even wanted a copy of Mary's check and my signature as the cutter of the jigsaw puzzle. We provided additional evidence in the form of a video my daughter Betty had taken—at least until she dropped the camera at the very end from sheer disbelief! A year and a half later, after seemingly endless overseas communications, the Guinness certificate arrived at Joy's house, and she lost no time driving over with the treasured document—confirming the highest price paid at a charitable auction for a hand-cut jigsaw puzzle.

A Painful Loss

The "empty nest" years that Mark and I enjoyed at the farm came to too soon an end when Mark fell gravely ill after a serious fall in 1995, during his 89th year, my 82nd and the 56th of our marriage. He somehow had a premonition that spring that death was not far off. One day as Mark and I walked through the lower pasture to our open field along the Concord River, a red-tailed hawk soared west overhead. As often before, we paused to sit on a big old oak stump and watch our dogs stir wild ducks from the brush near the shore. Words were not needed to share the love and memories of our years together. When the old farmhouse and barn came into view as we walked back, Mark put his arm around me and said quietly, "Honey, I think my time is near." I couldn't believe what I heard. I never imagined life without him and my voice failed to respond.

As I recall that day, I am reminded of the first time that we walked across River Road Farm, in 1946, when the land became ours. Ever the hunter, Mark spied a red-tailed hawk above the derelict old farmhouse and pronounced it a good omen. But the red-tailed hawk we saw that spring day in 1995 must have been a different kind of nature's omen.

Within a matter of months, Mark tripped and fell as we were leaving a friend's house. He struck his head. It was evident immediately that it was life-threatening, yet even as I listened to doctors at the hospital and saw his injury, I somehow couldn't believe he would be taken away from me. Yet to my sorrow that is what happened. Mark never recovered and died September 29th.

I was present when the undertaker, Charlie Dee Jr., arrived by hearse at the hospital. We had already engaged the Dee Funeral Home in anticipation of a memorial service at First Parish Church in Concord, a celebration of Mark's life. Before leaving the hospital, I asked Charlie (whose father was a close Rotary friend) if it might be possible to drive first to River Road Farm. I wanted the hearse to circle the fields, taking Mark on a final circuit of the land he so loved and protected for all of us. Charlie graciously agreed.

I didn't see it that day, but longtime friends Don and Lillian Stokes, who witnessed the procession, later wrote of what they saw. "As the hearse passed the pond on its way out," they wrote, "the red-tailed hawk was soaring right above in the deep blue sky."

Perhaps at that moment I was distracted by Charlie Dee's worries. As we drove slowly through the fields, our horses came forward of curiosity, as they do to any car driving through the pastures. This upset Charlie, but the slower he went, the closer they came, until, unnerved by their boldness, he made the mistake of coming to a complete stop. The horses then surrounded us and a few began to nuzzle the shiny black Cadillac hearse. Fear that they would leave scratches on the funeral home's new hearse seemed to throw Charlie into a state of shock. I jumped out to chase the horses away. This may explain why I did not see the red-tailed hawk that afternoon, but I am comforted to know it was there for Mark's final passage around River Road Farm.

The family at reunion in Georgetown, Maine, 1994. L to R: Betty, David and Ruth behind Mark and me.

Chapter 10

THE PADDOCK AND THE HILLTOP

The gardens I've nurtured and loved caring for during my 60-plus years at River Road Farm are yielding here and there to lawn. My stiff joints don't deter me, however, from occasionally planting or weeding, and I'm ever grateful for a little help. The daylily garden that Bob and Love Seawright planted here to grow *hemerocallis* and *hosta* plants for sale has long since become a pony paddock. This colorful and enduring chapter in the farm's history began when these long-time friends set out to make a business of their daylily hobby. They needed space to expand their bulb production and an unused acre of pasture right below our lawn proved to be the perfect solution. Thousands of brightly colored blossoms brightened our view of the pastures for many years until our friends moved their blossoming business to larger fields closer to their home. Many of their daylily varieties still flower here from late spring to early fall lending color to weathered fieldstone walls around the farm. Three of those varieties "Rachel's Hope," "Princess Sarah" and "Cynthia Page" were named for my granddaughters.

Occasionally, I still participate in the agility activities, but most of all I enjoy my short daily walks around the pasture nearest the house for my own exercise. Falling is sometimes a worry, but my trusty Golden, Tammy, is always ready to lend her strong shoulders as support should I need help getting up. My daughter's Welsh Corgis, who often stay here, and one or both cats, often trail along on these walks. They all stay close to me at the end of the pasture where I like to rest and contemplate beneath the wild apple trees.

As I look back across the open field, I appreciate how comfortably our friendly old house is shadowed by the maples we planted decades ago to replace elm trees eventually killed by the Dutch elm blight. At the end of the house, towering over all, a massive, 300-year-old black oak keeps watch. Built into a retaining wall below the lawn is one of two old stone wells. Beside it and still thriving is the witch hazel bush that my mother planted there when we moved here in 1946. "Always plant a witch hazel by a well," she said, "because it helps draw water." For that reason, she explained, dowsers traditionally use a forked witch hazel branch.

On the far side of the house, by the front door, is the oldest well, which once gave yeoman water service to the farm. It is now safely covered and resting on top is a large ceramic frog that our son, David, sent home from Vietnam as a souvenir of his wartime service there. Children still love to climb up and sit next to it. Spreading over the frog and stone well are the strong, graceful branches of our oldest maple, Grandpa's Maple, a welcoming landmark to all visitors to River Road Farm. As I look around on these daily walks, I realize how blessed I have been by the love and support of my friends and family, even now as the newest generation of great-grandchildren has come to this extraordinary place.

A Village's Warm Embrace

I've also been blessed by the town of Carlisle. When Mark and I moved to this community of just 400 residents in 1946, little did we dream of the warm welcome that lay ahead or the lasting friendships we would make. Neighbors Edie Booth and Peggy Grant welcomed us with flowering shrubs that still brighten the driveway every spring. Mabel Bates brought over fresh vegetables and occasional casseroles—a welcome supplement to the meals I could make on a coal-burning stove we used in the early days on the farm. I remember how Mabel's husband, Dick, carefully checked the fences before moving his cattle into pastures not grazed in a long while. When Mark and I temporarily supplied water to our neighbor Frank Biggi after his well went dry, he later thanked us with vegetables and freshly picked raspberries. Fortunately, many years before we bought the farm, Mason Garfield augmented the well water supply by persuading Concord to extend its water main along Monument Street to River Road Farm. That change was essential to the success of Mason's large dairy business, River Road Jerseys, which produced milk and cream from his carefully bred Jersey cows.

As newcomers, Mark and I were quickly drawn into town activities and committees, including service on, among others, the Carlisle Board of Public Welfare, the Carlisle Teachers' Association, the Cub Scouts, Garden Club and teaching Sunday school at the Unitarian church in town. We switched to First Parish in Concord, in our neighbor town, when my brother-in-law, Dana McLean Greeley, became its minister. With Ruth and Betty's devotion to ponies, there were the 4-H and pony clubs that Mardi Perry and I headed for several years. Goldwood Toby also kept me active in a local dog training group, which early on met in our barn.

Now and then I found time to gather wild Concord grapes for which Carlisle is noted. Only once was I chased off while doing so.

I thought all along that I was on town land, but I was actually picking on private property. The owner, Dr. Towle, quickly forgave me, however, when he learned that I was the lady who had introduced grape jelly venison stew to the men's supper group of our church, the Laymen's League of the First Religious Society of Carlisle. My reputation for this culinary innovation was an accident. After hunting trips, Mark occasionally contributed venison to the supper which I would cook into a stew. Once, as I was putting in the last touch of flavoring, someone appeared unannounced at the kitchen door to get my advice on breeding his Golden. Totally distracted by our talk, I dumped nearly an entire jar of grape jelly into the pot. There was no backing down. Mark took the concoction with him to the supper meeting and happily found his friends hungry enough to eat every bit. I learned to my surprise that they seemed to enjoy it, and years later Dick Bates told me that the group had hoped they would some time have another purple meal just like it.

A Sense of Pastures Drawing Closer

Getting older is ever present in the back of one's mind at any age, but the passing of years strikes hard when I run into grandmothers whose teeth Mark straightened as the first orthodontist in Concord. One slowly makes one's peace with this reality. It wasn't until I reached 93 that I started accepting the physical limitations that beset me with oncoming years. Lack of balance is a major inconvenience, requiring the use of a cane, not only for my stability but also to keep curious ponies at a distance when I am walking in their pastures. I hate to admit it, but I no longer drive after dark or for long distances. It seems that doing even familiar things takes more time, more caution, and I am more willing to accept offers of help from family, friends or strangers.

When I asked Alberta how I should approach discussion of these limitations with others, her answer was simple. "Just tell them you can't do anything about it." I wish it were that easy! I'm grateful to those who lend a hand grooming Tammy, shoveling snow from the front steps or doing other chores on my "honey-do" list that Mark used to refer to when I had a repair job for him. I still like the sense of freedom driving my car with a few dogs into the pasture and up to the circular field overlooking the river. In much the same way I insist I would still be riding or driving if my trusty steed, Telstar, were there to meet me at the front door and take me for a slow and gentle ride around the farm.

I remember some years back when my sister, Deborah, visited and she used to grab any strong arm available to help her navigate the unforgivingly high steps that were here when we moved in. One day she told me, quite frankly, that even though I was several years younger, we both needed a rail for easier access to and from the front door. We heeded the warning and before long a "Deb rail" was installed, so named to this day. If truth be told, Mark was also inspired to install other rails where most needed—a safety precaution for any household, whatever one's age.

The joy and challenge of growing older combines the gratitude of sharing so much with so many friends, but also in time the loss of them. Over the years I've learned that friendships should never be taken for granted. Recently, I lost a longtime friend who lived a mile up the road. I had known Ollie Lawrence since we went to high school together in Lexington. We were not especially close, but her death was still hard because it felt as though one more piece of my familiar world was gone. How often we drive by friends' houses, as I often drove past hers, and consider stopping to say hello—and then don't for lack of time or fear that it would be inconvenient. What a shame. I enjoy it when friends drop by, sharing stories over tea and keeping in touch with each other and the outside world. Sometimes we just repeat stories or laments of failing health, about which nothing is more healing than a good laugh. I've come to realize that we may be helping our visitors as much as they seek to help us. Just don't stop unannounced to talk about dogs while I'm making venison stew!

The Bounty of Many Years

Growing older has distinct advantages like becoming more patient and tolerant, wiser and more thankful, and learning to accept the kindness of others, developing a deeper appreciation of life itself, and knowing the value of a hardy sense of humor. I'm also more aware of the courtesies of those who offer an arm, pick up small dropped items or lift packages for me in the grocery store. Even people who wait in line patiently behind me as I poke for change in my purse display kindness. These small acts do matter. And the reciprocal is true. At the supermarket, I try to pronounce the names of the new Americans who bag my groceries or I inquire about their countries, and I enjoy it when they react smiling with pride. They seem to remember me, but perhaps only because I'm the lady who keeps asking them to keep the bags light. The same is true with the employees who wheel my shopping cart through the parking lot and load the bags in my car. Tammy often noses their outstretched hands, wagging her tail, surprising them because she never tries to jump out of the open tailgate.

Short-term memory does decline with age, it is true, but getting older physically does not have to mean getting older mentally. I like the story of the little girl worried about her great grandmother. "Grandma," she said tentatively, "you have wrinkles in your face and your hair is silver. Are you really old?"

"No, dear," replied the great grandmother. "I've just been young a lot of years."

I still enjoy new challenges and feel that there is much yet to be done—if I ever get the time. Chess puzzles still intrigue me and it would be helpful to learn to use a computer and the Internet. I love to spend time at the piano playing tunes that I learned in those chilly early morning hours in Lexington ninety years ago. Recently I tried my hand again at sculpting, using a metal armature that a well-known sculptor had designed for me years ago when I took a graduate course in sculpture at Harvard. The subject? A horse, of course, inspired by the beautiful, mid-1800s study in equine musculature created by Isidore Bonheur, brother of painter Rosa Bonheur. Working with clay is far more challenging than a soap sculpture, but well worth the effort.

Daily, I can hardly turn a corner at home without coming across some long-planned or unfinished task or craft project awaiting completion. There are piles of journals, letters, dog research materials and three-ring binders of photographs and clippings to sift and sort. There's a cabinet of colorful prints waiting to be cut and even orders for a few puzzles.

The key to staying young is not the latest drug or diet, said the writer Sam Ullman, but holding onto the excitement children have for the "joy and game of life" and their "sweet amazement" at the stars.

Age never interfered with my love of riding shown here one of my favorite Connemaras, Telstar, about 1982.

Pausing in the agility pasture with Tammy, my friend Donna Cutler and her beautiful Goldens, July 2008.

The Mystery of Wordless Communication

River Road Farm, as I observed in the beginning of this book, is quieter these days. Although the ponies do their share to keep the fields open, advancing pine trees, cedar, willow, ash, wild apple and buckthorn bushes intrude along the borders. No matter, for it is nature claiming her own. This magnificent place still speaks to me, just as it did that day I walked around it for the first time in 1946. Thankfully heeding the advice of Aunt Edith, I did give this old farm a chance, and I became happier for that with each passing year. As Carlisle and surrounding towns develop and change, I'm more aware of River Road Farm as a precious emblem of old New England—a gently rolling landscape of gardens, fields and ponds, with walls built of the field stones attesting to the labor the early settlers exerted in preparing and tilling the rocky soil. But its fields now mainly grow memories for me, as I am reminded when I find, tucked in here and there around the farm, herbs from Mother's famous garden in Lexington.

Undisturbed beneath the soil are the remains of many beloved animals, large and small, that shared their lives with us, offering years of unquestioned devotion, service, loyalty and kindness. What more beautiful place than these fields as a final resting place for them?

The loss of each horse or dog was difficult and healed only with the passage of time. The farm isn't haunted by their ghosts. It is brightened and rejoices in their spirits. It's hard to convey in ordinary words, but I see each of these faithful friends. They are alive and real to me. Their affection, honesty and devotion remain a living presence—and a standard to us all.

I recall now the day 62 years ago that my mother planted the witch hazel by the well. She seemed especially pleased, and I now wonder if she somehow looked into the future and knew in her heart that this old farm was the right spot for her high-spirited daughter, Father's "blue-eyed boy" who rode open trails out West and who loved freedom and the wind in her hair. Mothers are wise that way. She was happy for me, I know, even as she continued to treasure the 1930 photograph of that same girl on a grassy Montana hilltop, sitting astride her favorite Western mount, Scotty Grey, and looking out over the world below.

But enough of my memories. It's time to take my dogs for a walk and then get back to cutting a jigsaw puzzle.

Riding Scotty Grey, my 'favorite horse in the West,' on a Montana sheep ranch, 1930.

PUBLISHER'S NOTE

One of the best things about being a publisher is the opportunity it presents to meet and get to know the incredibly talented people who write the books we all enjoy so much. In our field of dog books, this means we have worked closely with famous judges, breeders, top trainers and behaviorists, and experts in nutrition and alternative medicines.

We have had a working relationship with Rachael Page Elliott for several years as the distributor of her DVDs, *Dogsteps* and *Canine Cineradiography*, and her book, *Dogsteps. A New Look*. We knew her by reputation, of course, but until we got involved with her on her memoirs we had no idea what an incredible life she has lived and what a talented person she is. And, even more remarkable, is the fact that at age 95 she is still active, pursuing a variety of interests, and doing research.

We just could not pass up the opportunity to reprint the following letter Pagey wrote to the editors of several dog magazines just two months ago concerning the 45 degree layback angle of a dog's scapula she discusses in this book. So for those of you who are wondering how this remarkable woman is doing, rest assured she is doing just fine!

May 17, 2008

Letter to the Editor:

The articles that have appeared in recent months in several dog magazines on the 45 degree layback angle of a dog's scapula with a 90 degree angle at the shoulder joint have concerned me. It is important to understand structure as it relates to gait and movement, however, it is equally significant to acknowledge the reality acquired with new information that refutes old beliefs that are structurally impossible. Unfortunately the theory of McDowell Lyon and others was accompanied by misleading diagrams that many of us accepted until the findings of new research came to light.

Thirty five years ago, at the time that the first edition of my book Dogsteps, Gait at a Glance *was published in 1973, I too had accepted without question Lyon's theory about shoulder structure and wondered, quite frankly, why we never found it. Curiosity led me to Harvard University's Museum of Comparative Zoology where special equipment, newly designed by Dr. Farish Jenkins, made possible the simultaneous x-raying and fluorescoping of dogs as they moved at varying speeds on a treadmill. In collaboration with the former President of the American Veterinary Medical Association, Dr. Edgar Tucker, my research on gait, structure and movement continued.*

Called cineradiography, or moving x-rays, this procedure revealed the action of bones and muscles as dogs moved at varying speeds on the treadmill making it possible to see the dog from the inside out. The study made it clear as to why a 45 degree layback of the scapula and 90 degree angle of the shoulder joint was an anatomical impossibility in normal canine structure.

With such an important revelation, it was imperative for me to get the information out to the dog world through a new edition of my book entitled The New Dogsteps. *Published by Howell Book House in 1983, this work was accompanied, shortly thereafter, with a video on the cineradiography research. In 2001, a third, updated edition titled* Dogsteps—A New Look, *continued the message with emphasis on sound structure, whatever we may ask of our dogs.*

I am reminded of one of my favorite quotations: "Information understood is knowledge. Knowledge understood is wisdom."

With scientific advancement and technology, new information became available to me and sharing this updated knowledge with canine friends was paramount. In the beginning, I did not have all the answers but the process of discovery has made me wiser.

In my forthcoming book, From Hoofbeats to Dogsteps, *I discuss in detail my research particularly on the front quarters about which much misunderstanding seems still to exist. I hope that your readers will learn that we are never to old to learn and will help me spread the word. At 95, I am still studying.*

Best wishes,

Rachel Page Elliott, Author

Dogsteps—A New Look

With Featherquest Sir Casper, 1992. Photograph by Jane B. Donahue.

Dog-gerel for Pagey in 1993

'Twas the winter of '89,
When Pagey declared, "Now's the time
To change my direction,
Review my collection
Of puzzles, perhaps earn a dime."

"I will not abandon old friends.
As my saw traces out all those bends,
I'll cut corgis and crabs,
Retrievers and labs,
And puzzlers will be at wits' ends."

Now she's spending such time on her puzzles
The dogs in the house have their muzzles
A bit out of joint.
They don't see the point
Of just some occasional nuzzles.

But puzzlers are filled with delight.
Such talent! Such grace! It sounds trite,
But she's one of a kind,
An unusual find,
Who loves to make friends at first sight!

With love and admiration from Anne D. Williams
(Professor at Bates College and author on puzzle history.)